ROUTLEDGE LIBRAR
HUMAN GEOGI

Volume 17

THE GEOGRAPHY
OF UNITED STATES POVERTY

THE GEOGRAPHY OF UNITED STATES POVERTY

Patterns of Deprivation, 1980–1990

WENDY SHAW

Routledge
Taylor & Francis Group

LONDON AND NEW YORK

First published in 1996 by Garland Publishing, Inc.

This edition first published in 2016
by Routledge
2 Park Square, Milton Park, Abingdon, Oxon OX14 4RN

and by Routledge
711 Third Avenue, New York, NY 10017

Routledge is an imprint of the Taylor & Francis Group, an informa business

British Library Cataloguing in Publication Data
A catalogue record for this book is available from the British Library

ISBN: 978-1-138-95340-6 (Set)
ISBN: 978-1-315-65887-2 (Set) (ebk)
ISBN: 978-1-138-95730-5 (Volume 17) (hbk)
ISBN: 978-1-315-66172-8 (Volume 17) (ebk)

Publisher's Note
The publisher has gone to great lengths to ensure the quality of this reprint but points out that some imperfections in the original copies may be apparent.

Disclaimer
The publisher has made every effort to trace copyright holders and would welcome correspondence from those they have been unable to trace.

THE GEOGRAPHY OF UNITED STATES POVERTY

Patterns of Deprivation, 1980–1990

WENDY SHAW

GARLAND PUBLISHING, Inc.
New York & London / 1996

Library of Congress Cataloging-in-Publication Data

Shaw, Wendy, 1951–
    The geography of United States poverty : patterns of depriva-
tion, 1980–1990 / Wendy Shaw.
        p.   cm. — (Children of poverty)
    Includes bibliographical references and index.
    ISBN 0-8153-2536-3 (alk. paper)
    1. Rural poor—United States. 2. Poverty—United States.
I. Title. II. Series.
HC110.P6S45 1996
362.5'0973'09048—dc20   96-42262

Printed on acid-free, 250-year-life paper
Manufactured in the United States of America

To my family; in memory of my father Ronald, to my mother Elsie, and my children Shaw Green, Fflur Green, Amanda Wallace, and Kassidy Zhou.

Contents

List of Tables   ix
List of Figures   xi
Preface   xv
Acknowledgments   xvii

I Introduction   3
    Background   3
    Unaddressed Questions   4
    Purpose and Study Questions   7
    Organization of the Book   8

II Theoretical Approaches Toward
   Explaining Poverty   9
    Introduction   9
    No Fault Theories   10
    Individual/Group Responsibility
        Theories   12
    Societal Responsibility Theories   13
    Government and Institutional
        Responsibility Theories   16
    Responsibility of the Economic
        System Theories   18
    The "Underclass" Theoretical
        Debate   23
    Shortcomings of Current Theory   28

III United States Poverty and Its
    Definition and Measurement   31
    Background   31
    Definition and Measurement   38

IV The Nature of Poverty and
    Characteristics of the Poor   53
        Introduction   53
        Dimensions of Poverty   53
        Characteristics of the Poor   56
        Spatial Variation in Poverty   60

V Analysis of Poverty's Distribution
    and Characteristics   67
        Introduction   67
        The Spatial Distribution of Poverty and the
            Identification of Poverty Cores   67
        The Concomitants of Poverty and
            Characteristics of the Poverty
            Population   81

VI A Spatially and Temporally Varying
    Model of Poverty   107
        Introduction   107
        Spatial Variation in the Concomitants
            of Poverty and Characteristics of
            the Poverty Population   107
        Temporal Variation in the Location of
            Poverty Cores and the Nature of
            Poverty, 1980 to 1990   168

VII Conclusions and Policy Implications   189
        Conclusions   189
        Policy Implications and Theoretical
            Insights   191
        Future Research   200

Appendix A   201

Appendix B   207

References   215

Index   241

Tables

1. Representation of States in the Poorest 150 Counties   76

2. Study Variables Utilized   82

3. Pearson Product Moment Correlation Matrix   88

4. Eigenvalues of the 1980 PCA factors   95

5. Communality Estimates for the Variables   96

6. Significant Factor Loadings and Interpreted Factor Meanings   97

7. Recomputed Urban and Rural Parameters from Appendix B2   144

8. Recomputed Regional Parameters from Appendix B3   149

9. Comparison of the 1980 and 1990 Income Index for the Poorest and Most Affluent Counties   176

10. Eigenvalues of the 1980/1990 PCA Factors   179

11. Communality Estimates for the Variables   179

12. Significant Factor Loadings and Interpreted Factor
    Meanings   180

13. Recomputed 1980 and 1990 Parameters from
    Appendix B4   184

14. Changes in Unemployment 1980 to 1990 for Selected
    Poverty Counties   186

Figures

1. Percentage of People in Poverty - 1959 to 1991   34

2. Millions of People in Poverty - 1959-1991   34

3. Spatial Variation in the Income Index in the U.S. - 1980   71

4. Spatial Variation in the Unadjusted Per Capita Income in the U.S. - 1980   73

5. Spatial Distribution of the Poorest 5 Percent of Counties - 1980   77

6. Poverty Density in the U.S. - 1980   78

7. Spatial Variation in the Black-White Duality in the U.S. - 1980   109

8. Spatial Variation in Non-Yuppieness in the U.S. - 1980   110

9. Spatial Variation of Extreme Age Population in the U.S. - 1980   111

10. Spatial Variation of the Urban Retail Economy in the U.S. - 1980   112

11. Spatial Variation of Employment Status in the U.S. - 1980   113

12. Spatial Variation of the Manufacturing/Blue Collar Economy in the U.S. - 1980   114

13. Spatial Variation of the Health Service Economy in the U.S. - 1980   115

14. Spatial Variation of Spanish Culture and Ethnicity in the U.S. - 1980   116

15. Spatial Variation of Native American Culture and Ethnicity in the U.S. - 1980   117

16. Spatial Variation of Health Care Availability in the U.S. - 1980   118

17. Spatial Variation of Population Growth in the U.S. - 1980   119

18. Spatial Variation of Labor Mobility in the U.S. - 1980   120

19. The Distribution of Urban and Rural Counties in the United States   121

20. Geographic Divisions of the Conterminous U.S.   122

21. The Effect of a Unit increase in the "Black-White Duality" on the Income Index, by Geographic Division   127

22. The Effect of a Unit increase in "Non-yuppiness" on the Income Index, by Geographic Division   128

23. The Effect of a Unit increase in "Extreme Age Population" on the Income Index, by Geographic Division   129

24. The Effect of a Unit increase in the "Urban Retail Economy" on the Income Index, by Geographic Division   130

25. The Effect of a Unit increase in "Employment Status" on the Income Index, by Geographic Division   131

26. The Effect of a Unit increase in the "Manufacturing Economy on the Income Index, by Geographic Division   132

27. The Effect of a Unit increase in the "Health Service Economy" on the Income Index, by Geographic Division   133

28. The Effect of a Unit increase in "Spanish Culture and Ethnicity" on the Income Index, by Geographic Division   134

29. The Effect of a Unit increase in "Native American Culture and Ethnicity" on the Income Index, by Geographic Division   135

30. The Effect of a Unit increase in "Health Care Availability" on the Income Index, by Geographic Division   136

31. The Effect of a Unit increase in "Population Growth" on the Income Index, by Geographic Division   137

32. The Effect of a Unit increase in "Labor Mobility" on the Income Index, by Geographic Division   138

33. Spatial Distribution of the Poorest 5 Percent of Counties - 1990   170

34. Spatial Variation in the Income Index in the U.S. - 1990   173

35. Spatial Variation in the Unadjusted Per Capita Income in the U.S. - 1990   175

Preface

This book was written in response to some glaring gaps in research and writing that have become apparent since the mid 1970s. It was in the early 1970s that interest in poverty in the United States was at its zenith and numerous written works concerning poverty, including its geography, were produced. Despite the persistence and worsening of poverty, since that time poverty's broad geographic dimension, its regional distribution and variation, has received comparatively little scrutiny. There has been little available detailed research concerning the geographic location of the poor in America and the analysis of regional variations in the extent and nature of poverty in the United states has been neglected. What recent research has been done on poverty within the United States has focused on national poverty levels, urban poverty, and the changing characteristics of the poor. Close attention has been given to particular segments of the population, and interest in poverty has largely focused on, and been confined to, the nature of urban poverty, its intraurban spatial dynamics, and the characteristics of the urban poverty population.

The poor outside large metropolitan areas have received minimal attention, yet the number of rural poor is substantial. The rural poor have become invisible, and rural poverty regions have been largely ignored within recent studies. Perhaps this is because there is an erroneous perception that the job of eradicating rural poverty has been substantially accomplished, and what is left is a mopping-up operation and a relatively relaxed wait for continued progress.

With this vacuum in terms of poverty research in mind this book seeks to address the following questions utilizing the spatial unit of the county.

1. What is the spatial distribution of the poor within the United States, and can distinct poverty regions be identified?
2. What are the concomitants of income poverty, and the general characteristics of the poverty population?

xv

3. How do identified concomitants of income poverty and characteristics of the poverty population vary in importance spatially?
4. Have the spatial distribution of poverty, the concomitants of income poverty, and characteristics of the poverty population changed from 1980 to 1990?

Chapter I provides an introduction to the issues, unaddressed questions, and research questions explored in this book. Chapter II goes on to explore the complex, and sometimes confusing, web of theory pertaining to poverty. Five categories of theory are presented: no fault theories, individual responsibility theories, societal responsibility theories, governmental and institutional responsibility theories, and responsibility of the economic system theories. Chapter III is divided into two main sections. The first section provides information on contemporary poverty in the United States as well has some historical background. The second section discusses the problems and complexities associated with defining and measuring poverty. Chapter IV details various dimensions of poverty and discusses the characteristics of the poverty population. Chapter V begins the research segment of the book. In this chapter the spatial distribution of poverty is analyzed and five clearly delineated cores of poverty identified. A model of poverty at the national level is developed using regression modeling. Chapter VI takes the national model of poverty developed in the previous chapter and explores it for both spatial and temporal variations. The spatial analyses focus on differences between urban and rural counties and variation by geographic region. The temporal analysis explores changes in the model of poverty from 1980 to 1990. Chapter VII offers some conclusions to be drawn from the research and discusses some possible policy implications.

It is hoped that this book will, in small measure, aid in understanding the nature and spatial distribution of poverty in the 1990s. This may be crucial as it appears likely that the welfare system faces major reconstruction with the establishment of a new political powerbase in Washington D.C. in 1995. The geography of poverty is a dimension of deprivation that cannot be ignored in an era of increasing demands for the most efficacious use of scarce tax dollars.

W.S.
Edwardsville, Illinois

Acknowledgments

I would like to thank Andrew J. Herod, Ikubolajeh B. Logan, Kavita K. Pandit, Clifton W. Pannell, and James O. Wheeler of the University of Georgia for their contributions and guidance during the production of this research. I am especially grateful to Kavita Pandit for her unfailing patience and support.

The Geography of United States Poverty

I

Introduction

BACKGROUND

The United States is among the most affluent countries in the world. Despite this general affluence, it is clear that poverty remains a widespread problem (Gugliotta 1994). As we move toward the twenty first century the problem of poverty appears to be growing worse. In 1994 some 39.7 million people in the United States lived below the "official" poverty line.[1] This represents the largest group of impoverished Americans since 1964, the year when an extensive campaign to address poverty was begun. Despite the fact that over 15 percent of the U.S. population lives in poverty, research on this topic has been surprisingly limited. However, a plethora of theory exists that, in some measure, seeks to explain the existence and growth of poverty amid general affluence within the United States.

Existing theory lays blame for poverty on a variety of causes. Some scholars contend that poverty is the natural result of variation in resource endowment or of the ups and downs of the business cycle. Thus, poverty is not a problem and should not be the target of intervention. A second group of theories sees poverty to be the result of individual actions and choice. The poor themselves are said to hold solutions to their own poverty and can escape impoverishment by changing their behavior. A third theoretical perspective asserts that the nature of society at large is the root cause of poverty. Discrimination on the basis of race, gender, and age are deeply entrenched in the fabric of American society; the result is rampant poverty among certain subsets of the population. A fourth body of theory suggests that the operation of government and major institutions leads to poverty for many Americans. Institutions are said to have failed or to be shelters of systematic discrimination. Government policy, especially in the area of welfare programs, is said to contribute, to rather than alleviate, the poverty problem. A final diverse group of theories sees poverty to be the result of the operation of the economic system. These theories do not see the operation of the economy as taking some natural form, but rather point to structural characteristics within the economic sphere which are the product of deliberate action. The shift from manufacturing to services, the move away from mass production toward flexible specialization, occupational segregation, the core-periphery sectoral nature of economic activity, change in the operation of the global economic system, and the general dynamic of capitalism are all posited as structural economic explanations of poverty.

It is clear that poverty is an ongoing social problem in the United States. It is also clear that an extensive and varied body of theory exists to explain this widespread poverty. However, studies concerning poverty have neglected important aspects of the phenomenon. In particular, spatial and temporal variations in poverty have received comparatively little attention and some important questions have not been addressed.

UNADDRESSED QUESTIONS

Much of the recent research on poverty within the United States has focused on national poverty levels, urban poverty, and the changing characteristics of the poor. Close attention has been given to particular segments of the population, such as children (Tobin 1990), female-headed households (Hoffman 1991; Renwick and Bergmann 1993), and the working poor (Levitan and Shapiro 1987). A spirited and ongoing debate within academia concerning the nature of the "urban underclass" has developed (Wilson 1980; Murray 1984; Wilson 1987; Hughes 1990; Sheppard 1990; Eggers and Massey 1992). Interest in poverty has largely focused on, and been confined to, the nature of urban poverty, its intraurban spatial dynamics, and the characteristics of the urban poverty population.

Poverty's broader geographic dimension, its regional distribution and variation, has received comparatively little scrutiny. Since the early 1970s (Brunn and Wheeler 1971; Morrill and Wohlenberg 1971), there has been little available detailed research concerning the geographic location of the poor in America. Despite empirical evidence of its existence, the analysis of regional variations in the extent and nature of poverty in the United states has been neglected.

Discussion has been largely limited to broad generalizations contrasting the rich North and poor South. Similarly it has been acknowledged within the literature that poverty may differ in nature between regions such as the Mississippi Delta and Appalachia but systematic research has been lacking. Poverty is often treated as a monolithic aspatial phenomenon, or at best geographic variations are treated as of only secondary and peripheral interest.

In addition to spatial variations in poverty, it also seems reasonable to conjecture that there have been significant temporal changes during the 1980s. The extent and nature of poverty may well have been impacted by the political climate of the 1980s as well as by structural changes within the national and global economies. As the 1990 census data become available, it is important to explore possible temporal changes in United States poverty.

This lack of attention to the aforementioned aspects of poverty may be attributable to various causes. Cities are the focus of modern American life, and urban poverty has captured media and hence public attention. Inner city poverty is often portrayed in lurid and dramatic ways, and the inner-city poor are stereotyped as a group apart from society that threatens the fabric of life and the validity of the American dream (Wacquant and Wilson 1989).

Studies of poverty's broader geographic dimensions offer no such drama or threat. The poor in regions outside the city live in areas that are backwaters both in actuality and in the psyche of most Americans. The average American may drive through Appalachia or the Southern Coastal Plain and give these areas passing cursory attention, but the poor there leave no etched images and appear to offer no social threat. Another possible reason for lack of attention to the broader geography of poverty may be the perception that the low levels of American poverty experienced in the mid 1970s meant that the extensive anti-poverty programs of the 1960s had been largely successful. The job substantially done, what was left for the 1980s was a mopping-up operation and a relatively relaxed wait for continued progress. While many may disagree that poverty has been largely eradicated, public opinion exerts a powerful influence on research directions. Yet another factor constraining poverty research again rests in public opinion and the conservative political climate of the 1980s and 1990s. The public perception seems to have developed that welfare support is extensive and jobs are available for those who are willing to work. Given this, there is no reason for anyone to be poor unless it is through personal choice or failings. The poor choose leisure or crime over work, become single parents because of promiscuity, and abuse drugs and alcohol. Within the political sphere numerous calls have been made to curtail welfare and other social spending that supports America's most impoverished citizens. Suggestions have been made to link work to welfare payments and to more closely monitor and administer existing programs to combat fraud. Understanding the nature and spatial distribution of poverty in the 1990s is especially crucial as it appears likely that the welfare system faces deep cuts and major reconstruction with the establishment of a new political powerbase in Washington in 1995. Also given this public and political climate, poverty research is not likely to receive much support and a search for structural causes of poverty, and their geographic manifestations, may be seen as futile.

This neglect of the broad geographic dimensions and variations within United States poverty has meant that the poor outside large metropolitan areas have received minimal attention. This void has serious consequences. Little is being done to improve the life of the non-metropolitan poor or to address their problems, and no theory is offered to guide current and future social programs. While urban poverty may affect many people, the majority of people in American urban centers are relatively affluent.

In contrast, there are rural areas where a large proportion of the total population exist in poverty. The number of rural poor is substantial, yet they have become invisible, and rural poverty regions have been largely ignored within recent studies. The links between poverty and place were the subject of discussion in the 1970s, but limited research resulted. Little is known about connections between poverty and local social contexts; geographers have seemed unwilling to go beyond the identification of the spatial distribution of deprivation.

Analysis of spatial and temporal variations in United States poverty promises to broaden theoretical understanding of the phenomenon of poverty. While existing theories have been postulated as broad explanations of the spatial concentration, nature, and dynamics of inner-city urban poverty, little similar attempt has been made with regard to rural poverty. Current theory has not attempted to embrace or explain poverty and its spatial concentration in rural America or to draw comparison with urban poverty areas. In addition to the lack of broad theory concerning rural poverty, there is also a lack of recognition of the possible importance of regional or temporal contexts that may contribute substantially to the causes of poverty. Existing theories provide generalized explanations of poverty and tend to be both spatially and temporally invariant. However, the factors that contribute to poverty may not be constant over space or time.

Investigation of the more localized character of sub-national regions may help to explain the spatial patterns of poverty and aid in understanding its persistence in some geographic areas. That intransigent regions of poverty seemed to persist even in the 1970s, despite a concerted effort to alleviate their problems, would seem to lend credence to the idea that unaddressed underlying problems and dynamics contribute to, and sustain, the deprivation being experienced.

PURPOSE AND STUDY QUESTIONS

The purpose of the study presented here was to investigate regional patterns of poverty in the United States, with a focus on spatial and temporal variations in the level and nature of American poverty and characteristics of the population who are impoverished. Four specific questions were addressed:

1. What is the spatial distribution of the poor within the
United States, and can distinct poverty regions be identified?
2. What are the concomitants of income poverty, and the
general characteristics of the poverty population?
3. How do identified concomitants of income poverty and
characteristics of the poverty population vary in importance
spatially?
4. Have the spatial distribution of poverty, the concomitants
of income poverty, and characteristics of the poverty
population changed from 1980 to 1990?

Any spatial and temporal variations in poverty may have important policy implications for identified poverty regions. Policy suggestions that emanate from the results of the questions posed, may be important in the search for ways to alleviate poverty. In addition, in an era of demands for budget cuts and more efficient use of tax dollars, policy suggestions may aid in the accomplishments of these goals. Thus, discussion of possible policy implications is an additional goal of this study.

The conterminous United States is the area focused upon in order to explore the spatial distribution of poverty and identify spatial cores of deprivation, and also to investigate the national concomitants of income poverty. Alaska and Hawaii are excluded for a variety of reasons. Both are geographically, socially and economically isolated from the continental United States. In addition, both states have only relatively recently become part of the United States and have populations that are demographically distinct.

County level data for 1980 and 1990, available from the United States Bureau of the Census, are utilized in this research. Variables are chosen that represent the socioeconomic dimensions of poverty and characteristics of the poverty population. Selection of the variables used was based on the extensive literature available pertaining to United States poverty.

ORGANIZATION OF THE BOOK

The remainder of the book is organized as follows. Chapter II details the various theoretical viewpoints pertaining to poverty that abound within the literature. Chapter III provides an overview of the historical and contemporary levels of United States poverty. This chapter also addresses crucial questions concerning both the definition and measurement of poverty. Chapter IV presents the major dimensions of poverty and provides a view of the nature of United States poverty and general characteristics of the poor. Chapter V is devoted to an analysis of the distribution of poverty and identification of poverty cores as well as to a model of poverty for the nation as a whole. The data, methodology used, results, and a discussion pertaining to these topics are included. Chapter VI provides a discussion of the data, methodology, and results of both the spatially and temporally varying models of poverty. The final chapter, Chapter VII, presents a brief summary of some major conclusions arrived at in the research. In addition some policy implications for each identified poverty core are identified and some suggestions for future research offered.

NOTES

1 The official poverty line is defined by a family income threshold that is specified by family size, age of the householder, and the number of related children under 18. Persons in households with incomes below the poverty line are characterized as poor.

II

Theoretical Approaches Toward Explaining Poverty

INTRODUCTION

Geographers have long recognized that social phenomena vary spatially, and have assiduously worked to point out the spatial patterns involved. However, Peet (1972) contends that geographic researchers have been loath to take any position concerning the root causes of poverty. Perhaps one factor in this shortcoming within geography is the lack of clear comprehensive theory concerning poverty. A hotch-potch of theory exists. Within each discipline theory is labelled and described differently, and the crucial issues identified vary. Not only is theory fragmented by discipline, but it is also divided by level. Theories have been put forth at the macro-, meso-, and micro-levels but researchers have been unable to draw the threads together (Oyen 1992). Problems of theorizing poverty are understandable. Poverty is not a neatly described and discrete phenomenon, but rather is a label for a varying cluster of consequences which are displayed from the individual to the societal level (Oyen 1992).

Existing theories of poverty, although confused, have been categorized and classified in various ways in an attempt to clarify the broad dimensions of theory. Auletta (1982) identifies a simple theoretical dichotomy with the causes of poverty being laid either on individuals or society. Similarly Ropers (1991) sees theories of poverty as divided between those who blame the victim and those who blame the system, and Kodras (1992) between those which point to individual failure and those which pinpoint market failure. Peet (1972) divides theory in terms of control; the first group of theories contend that the poor retain control of their economic fate while the second asserts that poverty is a result of economic and/or social circumstance that is beyond the control of those who are impoverished. Other scholars have suggested more complex theoretical categories. Wachtel (1974) takes a political view; theories are either conservative, liberal, or radical. Smith (1987) attempts a more geographic classification where poverty is the product of local conditions, the role in the spatial network of places, or of the political economy of place. Dhillon and Howie (1986) suggest that there are five approaches to poverty; individual/genetic, cultural, opportunity, mal-distribution, and scarce resources. Holman (1978) characterizes explanations of social deprivation as falling into the broad areas of individual, cultural, institutional, and structural.

Scrutiny of the vast and varied literature and research pertaining to poverty suggests a somewhat different categorization of theoretical perspectives. Poverty is usually (but not always) viewed as an undesirable phenomenon. Thus, when poverty is discussed it is usually couched in terms of 'fault'. When assigning blame for poverty, five general culprits seem to emerge. One group of theories see poverty as no one's fault, but rather the product of nature or uncontrollable social, political, or economic forces. A second group of theories lays blame on individuals or the poor themselves as a group. From this perspective the poor largely cause their own poverty. A third group of theories see the causes of poverty to be the fault of the general culture and attitudes of society. A fourth theoretical perspective points to government and institutional culpability in the creation and maintenance of poverty. Finally, a varied group of theories can be generally described as seeing poverty as rooted in the operation of the economic system.

NO FAULT THEORIES

Within this category fall theories which view poverty as a result of general economic health (Greenstein 1985; Sawhill 1989; Solow 1990; Tobin 1990). In addition theories that point to variation in resource endowment or the natural environment as the cause of inequality fall within this theoretical perspective.

Poor economic health is said to be characterized by economic stagnation, underemployment, low wages, and increasing unemployment (McCormick 1988; Solow 1990; Eggers and Massey 1991). Such problems may be closely tied to the natural ups and downs of the business cycle or be the result of low levels of development within a region (Mattera 1990). In any event solutions are clear and are to be found in the free operation of the market and in the stimulation of economic growth (Solow 1990). Those who advocate growth and development as solutions to poverty are numerous (Peterson 1981; Hormats 1984) and acceptance of this idea underlies much of the unbridled pursuit of growth in American cities and elsewhere (Wolfe 1981; Hughes 1991). Growth leads to the greatest good for the greatest number of people (Sonstelic and Portney 1975; Peterson, 1981; Logan and Molotch 1987). Resources must be mobilized in the pursuit of growth in order to improve conditions for all (Sonstelic and Portney 1975; Lund 1976; Fosler and Berger 1982). Once economic growth is established, a city or region increases its competitiveness and attracts even more investment, and thus becomes increasingly prosperous (Andrews 1962; Fosler and Berger 1982; Noyelle and Stanback 1984; McCombie 1988a; 1988b). Those who advocate this macro-economic growth approach contend that eradication of poverty will come naturally through "trickle down" of economic benefit gained through general economic growth (Velasquez 1987). Many argue that such economic growth is best facilitated by helping markets operate more efficiently, and that market efficiency is reduced by income supports and welfare programs (Tootle 1989). Critics of the economic growth solution point to the fact that there appears to be a diminishing response of poverty to economic progress (Tobin 1990).

Viewing inequality and poverty as a product of the natural environment or resource endowment and distribution has a long history. Environmental determinists proposed that varying levels of development and well-being were a direct result of the effect of the natural environment on human activity. Although Possibilism rejected such determinism this perspective still acknowledges the constraining and enabling effects of the environment (Rubenstein 1989). Many classical economists point to comparative advantage in terms of resources and the resultant division of labor as natural (Ohlin 1933; Linder and Kindleberger 1982). Moreover, inequality is justified and in everyone's eventual mutual interest (Reitsma and Kleinpenning 1989). Theories which point to resource differentiation as the root of poverty continue to be supported (Smith 1987). Both Deavers (1980) and Macnicol (1987) cite part of the problem of poverty as lack of local resources and opportunity. Smith (1990) indicates that some places are dying as a result of multifaceted disadvantage and should be abandoned. For individuals who live in regions which lack environmental or resource advantages the only logical recourse is to leave (Smith 1990).

INDIVIDUAL/GROUP RESPONSIBILITY THEORIES

Theories of poverty which fall in this category lay causality for the condition squarely with poor individuals and groups, and the emphasis is on behavior (Rickets and Sawhill 1988). Individuals are said to have either flawed characters or to display characteristics that are responsible for their condition. Within this general framework poverty is a direct result of individual decisions or is a product of a culture of poverty. In either event the poor have no one to blame but themselves (Ropers 1991).

These views that hold the individual responsible have a long history, and were prevalent in Britain in the nineteenth century. The poor were portrayed as idle and immoral. Tightening the poor laws in this period was designed to alleviate the burden on the productive portion of society, and to be punitive against the poor (Katz 1983). Ideas of individual responsibility have been periodically revived (Macnicol 1987).

The 1960s was a decade in which poverty in the United States was blamed on the victim. It was during this decade that ideas concerning a culture of poverty were much in vogue. The emphasis was on the family and on the transmission across generations of the behaviors, attitudes, and values that perpetuate and reinforce poverty. The poor, by adhering to their "culture of poverty", followed a lifestyle that was at the basis of their own impoverishment. According to this perspective the poor clearly must enter mainstream culture to have a chance for economic success (Dobelstein 1987; Dudenhefer 1993).

The flawed character and culture of poverty views are by no means a thing of the past (Jordan et al. 1992). In fact Schiller (1980) identifies these views as the most popular, and reinforced by statistical profiles of the poverty population. Individual explanations of poverty center on three types of problem within people that can result in poverty (Holman 1978). The first set of deficiencies are genetic and may be manifest in low intelligence, instability, or mental illness. The second group of problems revolve around economic decision making. People are poor because they are lazy and prefer leisure over work (Galloway and Vedder 1985). The poor are said to deliberately avoid work and to be unwilling to take the many jobs that are available (Mead 1989; 1992). A third perspective points to just a few people as being at the core of the problem. These people are social deviants who cause chaos in the community with their criminal activity and who live in profoundly dysfunctional households (Holman 1978; DiLulio 1989). In both the second and third cases individuals have chosen to reject the norms of society (Moynihan 1989; Sawhill 1989), and solutions are to found in enforcement of behavioral norms (Mead 1989; 1992).

Moving beyond individual causality the poor as a group are often said to create and perpetuate their own condition (Ryan 1974). In the 1960s Oscar Lewis (1969) suggested that a contemporary subculture of poverty exists which has its own structure, logic, and rationale. This subculture is both a reaction and an adaptation on the part of the poor and once in existence serves to perpetuate poverty across generations. This subculture, Lewis contended, exists on the individual, family, and community levels (Lewis 1969). Many researchers embraced Lewis' concept of a culture of poverty finding in it an explanation for the inter-generational transmission of poverty and chronic poverty within certain households. Families especially are identified as units within which undesirable behavior is created and sustained (Holman 1978; Moynihan 1989; Mead 1992).

Many question both individual responsibility and culture of poverty theories (Valentine 1968; Leacock 1971; Katz 1989), seeing them as serving ideological positions rather than representing reality. They point out that there are many people who work diligently and follow all society's rules yet remain poor (Schwarz and Volgy 1992). However, these theories continue to be widely voiced (Rossi and Blum 1969; Ryan 1974; Macnicol 1987; Wacquant and Wilson 1989; Williams 1992).

SOCIETAL RESPONSIBILITY THEORIES

The culture and attitudes of society are often cited as major causes of poverty among certain segments of the population. Discrimination on the basis of race, age, and gender erects artificial barriers between workers and jobs, restricting upward mobility (Schiller 1980; Fitchen 1981). Negative judgements and stereotypes of the poor by the larger society may also serve to restrict the choices of impoverished Americans (Fitchen 1981). Researchers such as Granovetter (1985; 1986) see economic action as socially embedded. Job searches and job opportunities are circumscribed by the structure of social relations (Granovetter 1985; 1986). In particular the sexism and racism which pervade society are often seen to be the underlying cause of the high rates of poverty among women and minority groups like

African Americans (Sawhill 1989; Wacquant and Wilson 1989). Poverty rates experienced within the elderly population are also said to be impacted by attitudes prevalent in society.

Sexism

In American society women face discrimination in various forms that influence their poverty status (Beeghley 1983). Women must earn a living in a discriminatory labor market. Women are highly concentrated in just a few occupations (Jones 1990; Nord and Sheets 1992); two thirds of all working women are employed in just ten occupations. The jobs that women do are often low-paid, and the result is that on average women's income is only 60 percent of men's. As well as being low-paid, women are more often part-time or temporary workers than are men. They also suffer from higher unemployment than do their male counterparts. The result of women's position in the labor market goes beyond just low income. The occupations in which women are concentrated, and part-time or temporary workers may not enjoy benefits such as health insurance or retirement plans (Jones 1990).

Women have more constraints than do men that may hinder their efforts to break out of poverty. Our culture and society, as well as many women themselves, still expect women to retain the roles of mother and homemaker. Household "duties" and family priorities may limit the kind and location of employment sought, as well as the hours worked. The nature of job contacts may also limit employment opportunity for women. Informal, personal, and family-based contacts in a quite narrowly circumscribed local area are common for women. Job information flows channel women into low-paid, traditionally female, occupations (Hanson and Pratt 1991). In addition, divorce impacts women differently than men. Frequently, women continue to bear the brunt of responsibility for any children while men are freed from family constraints (Beeghley 1983).

Some view women to be in a doubly disadvantaged position. The job segregation and discrimination faced by women serves the capitalist system and capitalists to their advantage. Channeling women into certain occupations and erecting barriers against entering others, guarantees a labor force with limited options who must accept low pay. Underlying this differential exploitation of women is a traditionally patriarchal system that has existed through much of human history and which places females in a subordinate position compared to males. This system continues to benefit males and so is maintained by men, but is used and exacerbated by capitalism for its own advantage (Hartman 1976; Dear and Wolch 1989).

Racism

Minority racial and ethnic groups also face multidimensional discrimination that is deeply ingrained in society. Many point to racial discrimination as a root cause of African American poverty (Duncan 1969; Ryan 1974; Auletta 1982). African Americans face continued racial segregation which is historically based. As is the case with gender, race is used as the basis for a divided labor market (Rossi and Blum 1969; Singh 1991; Nord and Sheets 1992). African Americans, if they can find work at all, are increasingly segregated by industry and plant and are pushed into the worst jobs. African Americans and other minorities may be excluded from various occupations or work places. Both job screening and job recruitment practices frequently contribute to the employment problems of racial minorities (Beeghley 1983). Employers often have negative stereotypical perceptions of African Americans as potential workers (Kirschenman and Neckerman 1991). In addition, social custom and pressure often results in segregation in such things as housing and public facilities (Frey 1993; Katz 1993).

Again, racial divisions are said to have existed prior to America embracing the capitalist system, but were subsequently used by it. The result is that black Americans, even though working, are likely to earn less than are white Americans (Bonacich 1976; Wilson 1980; Beeghley 1983; Wilson 1987). Organized labor must share its portion of responsibility for poverty among African Americans. Black Americans were often systematically denied union membership, and since such membership was a condition of employment in some industries this accentuated occupational segregation. In addition, those excluded from participation in unions could not enjoy negotiated benefits. Thus, union activity is partially responsible for racial occupational segregation and the concentration of African Americans in the least attractive jobs (Oppenheimer 1985).

The Elderly

It is clear that in the recent past the elderly were a severely impoverished group in the United States. However, in recent years poverty among older Americans has been greatly reduced largely because society had the will to do so. This decrease in elderly poor was due partially to Social Security payments and Medicare (Haveman 1987).

America's elderly population receive a large proportion of governmental social spending. More social programs are age designated than are means tested, social security providing one major example (Harrington 1987). The reasons for this substantial social spending on the elderly are twofold. As the population has aged the elderly have become an increasingly powerful special interest group who some say receive preferential treatment. In addition, a crucial factor is that in the public perception the old are more clearly a public responsibility than are other groups (Axinn and Stern 1988). While some of the poor may be thought by society as undeserving, impoverished older Americans are more likely to be regarded as deserving of financial support (Katz 1989; Kodras and Jones 1990; Bremner 1992; Katz 1993). So far the elderly have been largely excluded from a growing sentiment against governmental action aimed at redistribution of income (Thurow 1982).

GOVERNMENT AND INSTITUTIONAL RESPONSIBILITY THEORIES

Another group of theories point to the operation of government and of related institutions as at the root of poverty (Deavers 1980; Nathan 1989; Solow 1990; Schwarz and Volgy 1992). Institutions are said by some to have failed, by others to be bastions of discrimination, and by some to act deliberately to maintain an unequal society (Rossi and Blum 1969; Holman 1978; Dhillon and Howie 1986). The State and government is seen by some to function primarily to maintain ideological hegemony and to protect the social, economic and political status quo (Carnoy 1984; Cox 1989; Laws 1989; Wolch 1989; Miliband 1990). In addition, government policy, especially social welfare programs, are identified as being either inadequate or too extensive (McLanahan 1989; Rossi and Wright 1989; Wilson 1989; Bremner 1992).

Ryan (1974) identifies the root problem of poverty to be pervasive institutionalized discrimination. Compensatory programs in such things as education and health programs can only alleviate symptoms but do nothing to restructure inherently discriminatory institutions.

Failure in the major institutions of society means failure to provide such things as a sufficient education and health care, and also structural bridges to the community at large (Fitchen 1981; Tobin 1990). The poor may lack such things as sufficient human capital but this is a result of the failure of society to provide education that is sufficient, appropriate, and of high quality. Lack of human capital which characterizes the poor is not a product of individual choice that fails to take advantage of available opportunity.

Researchers such as Roby (1974) and Piven and Cloward (1974; 1982) charge both institutions and government with acting to maintain an unequal social order. Social programs do nothing more than spend enough to pacify the poor. In fact programs can be punitive, serving to deter people from seeking relief and making low wage menial jobs a more attractive choice (Piven and Cloward 1974; Roby 1974; Carson 1991; Ropers 1991). Poverty and inequality are functional, providing benefits for some powerful groups. The poor provide low wage labor, keep prices of goods and services down by providing cheap labor, provide jobs with government agencies by such activities as their use of welfare, buy goods and services others do not want, and absorb economic and political costs. Given these benefits, the existing inequality is protected and public assistance is designed to keep people poor and compliant (Beeghley 1983; Johnston 1990; Ropers 1991; Oyen 1992).

A major theme that runs through conservative explanations of poverty is that government welfare programs are too extensive. The level of welfare payments is said to encourage and enable behaviors that result in poverty (Dudley 1988; Sawhill 1989; Singh 1991). The poor choose welfare and leisure as preferable to work and become dependent on transfer payments (Auletta 1982; Mead 1989). In addition, welfare claims are perceived often to be fraudulent (Beeghley 1983; Harrington 1984). There is a growing call by those who hold such views to link the receipt of government payments to work (Sawhill 1989). The able bodied should be obligated to do designated work. In this way, the cycle of dependency can be broken, mainstream cultural values enforced, and the poor forced to pursue their own economic interests rather than burdening the productive segment of society (Mead 1992).

The conservative view of social programs is hotly contested by liberal scholars. Social programs, including welfare payments, do not go far enough to help the needy (Harrington 1984; Hoffman 1991; Anglin and Holcomb 1992). The programs that do exist are poorly organized, articulated, and implemented and are reactive rather than proactive. They make untested assumptions about who the poor are, what their needs are, and may even create serious problems for impoverished Americans (Wachtel 1974; Higgins 1978; Kodras and

Jones 1990; Carson 1991; Kodras 1992; Williams 1992). In addition, most transfer payments benefit the rich not the poor. Even the tax system works to the advantage of the affluent and to the detriment of the impoverished (Harrington 1984; Ropers 1991). Put succinctly, there is welfare for the rich and a strict market economy for the poor (Ford 1973).

RESPONSIBILITY OF THE ECONOMIC SYSTEM THEORIES

Theories which identify structural causes of poverty within the economic system are varied (Dudley 1988; Knox 1991; Anglin and Holcomb 1992; Davis and Herbert 1993). They include theories that see poverty to be the result of the rise of services, the location of jobs, change in the production paradigm, occupational segregation, the sectoral nature of economic activity, change in the global economy, and the general nature and dynamic of the capitalist economic system. These structural theories of poverty are often obviously closely and intricately interconnected.

The decline in manufacturing and the shift toward services has been the focus of many explanations of poverty in recent years (Harrington 1984; DiLulio 1989; Sawhill 1989; Knox 1991; Williams 1991). Since the early 1970s an increasing percentage of workers have been employed within the broadly defined service sector (Noyelle and Stanback 1984; Bednarzik 1990; Nord and Sheets 1992), and almost all new jobs are within this sector (Ginsberg, Noyelle, and Stanback 1986). This structural change in the economy has meant that the location of jobs has changed. Many manufacturing jobs once concentrated in the central cities of the northeast no longer exist (Stanback and Noyelle 1982; Noyelle and Stanback 1984; Noyelle 1987; Noyelle and Stanback 1990). Instead they have been replaced by service sector jobs that are more likely to be found in the suburbs and in the south and west "sunbelt" regions. Thus, one result of the shift to services has been unemployment for those people who are unable to relocate (Wilson 1980; 1987). In addition, this change in the nature of employment opportunities has meant that jobs are often less well paid (Noyelle and Stanback 1984; Noyelle 1987; Appelbaum and Albin 1990; Jones and Kodras 1990). Service sector jobs are frequently very poorly paid in contrast to jobs within manufacturing, a situation which may explain the growth of the working poor. The mid-level incomes previously to be found within manufacturing are a rarity within services (Sassen 1988; Tootle 1989; Nord and Sheets 1992). Thus, poverty may not only reflect lack of work but also the quality of jobs available (Massey 1984). Even in the manufacturing jobs that do remain, change has occurred. One such change involves robotization or computer-based automation that displaces workers (Harrington 1987; Block 1990). One more result of the shift toward services has been a decline in demand for poorly educated unskilled workers. Even low level service jobs often require relatively sophisticated literacy skills (Kasarda 1989; Rossi and Wright 1989; Sawhill 1989). It is this shift to services in the U.S. economy that has often been cited as the root cause of concentration and intensification of poverty among minority racial groups in the last twenty years (Wilson 1980; 1987; Tienda 1989; Nord and Sheets 1992).

A change in the manufacturing paradigm from mass production to flexible production has occurred and is also posited as one reason for the nature and extent of poverty in the United States (Schoenberger 1988; Kodras 1992). This fundamental structural change has created confusion within the economy since late in the 1970s when it began. The breakdown of mass production and rise of flexible production and specialization has meant increased social embeddedness of the economy. The economic and social spheres have become increasingly intertwined and effective economic action more dependent on a complex network of social connections. Extensive social linkages characterize the new paradigm. Small firms exist on short-term subcontracts and a group of workers often come together on a temporary basis (Piore and Sabel 1984). The era where a single breadwinner could work in a traditional mass production industry and do the same job for a lifetime is rapidly dissolving (Jones and Kodras 1990). Employment has become much less secure, and employees must have multiple skills and be relatively highly educated (Piore and Sabel 1984). Flexible production has brought a new set of problems that existing social programs are ill equipped to deal with. This failure is likely to result in deepening of poverty for some people, despite overall affluence (Kodras 1992). Cox (1989) and Oppenheimer (1985) also identify a failure within organized labor to respond to the changing nature of production and employment.

Occupational segregation is another structural factor within the economy which is sometimes identified as a major cause of poverty. This segregation appears to be associated with the sexism and racism discussed earlier (Harrington 1984). Occupational segregation operates at the conjunction between social prejudices and the desires of the economic system. As has been previously mentioned, it is often women and minority racial and ethnic groups that suffer the most negative consequences (Tienda 1989; Tienda et al. 1987; Jones 1990; Hanson and Pratt 1991; Jordon 1992; Nord and Sheets 1992). These groups are cited as lacking the right kind of contacts, having limited job information, and suffering from household and residential constraints (Hanson and Pratt 1991). The labor market is deeply discriminatory. Whites, particularly white males, dominate those occupations that offer the highest economic rewards, the most security, the most prestige, and the highest status. In contrast women and African Americans are highly concentrated in low-level, low wage, unpleasant occupations which are also the least sheltered (Jones 1990; Hanson and Pratt 1991; Jordon 1992; Nord and Sheets 1992). A vivid example of the occupational segregation of women is given by Jones and Kodras (1990) when they show that 99 percent of all secretaries are female.

The sectoral nature of economic activity is another structural dimension that can profoundly affect economic outcomes for individuals. A central concept is that a dual economy exists (Averitt 1968; Stokes and Anderson 1990). The theory of a dual economy divides firms or industries into core and periphery sectors. The core sector is characterized by large vertically integrated organizations. Such organizations produce their own inputs and operate in diverse non-competitive product markets (Tigges 1992). Core sector firms or industries enjoy many advantages, including shelter from competition, and can pass on their advantage to workers employed in the core sector (Hodson 1984). In contrast, peripheral sector organizations tend to be small, and must operate in the competitive market place. They must bear the full brunt of market forces, and have no advantage to pass on to their workers (Hodson 1984; Tigges 1992). Associated with the dual economic sectors identified, is a dual labor market. Gordon et al. (1982) identify a primary labor market associated with the core economic sector. Within the primary labor market individuals may improve their position, and capitalize on skills. Such improvements often come within the sheltered environment of the internal labor market. Employees may spend their working life within one firm or industry, improving their economic position by career moves up the internal ladder. Within the secondary labor market, associated with the peripheral economic sector, such internal career trajectories are not a viable option. The secondary labor market is characterized by dead-end jobs, and by a structure that can offer no reward for improved skill or experience (Gordon et al. 1982; Tigges 1992).

Some researchers contend that, to understand economic conditions in any area, it is necessary to view the world as a global economic system (Wallerstein 1974; 1979; Frank 1987; Peet 1987; Cox 1993). Change in the global economy is another structural transformation cited as a cause of contemporary poverty in the United States (Sanderson 1985; Gordon et al. 1982; Gordon 1988; Cox 1993). Since colonial times the international division of labor has meant that today's less developed or peripheral countries have exported raw materials while the more developed core countries have dominated manufacturing. Dominance of manufacturing and the global economy by countries such as the United States has meant affluence for these countries (Frank 1979; Lipietz 1986).

In recent decades a "New International Division of Labor" has emerged (Frobel et al. 1977; Frank 1987). This change in the economic world order involves the diffusion of manufacturing industries to countries which previously specialized in the export of raw materials. This shift in the location of manufacturing has meant that the United States now faces industrial competition and deindustrialization (Deavers 1980; Bluestone and Harrison 1982; Harrington 1984; Massey 1984). This change in the international division of labor is largely the result of actions by the State and also by multinational corporations. Multinational corporations control international flows of capital and have been instrumental in the development of what is often characterized as a global factory (Barnet and Muller 1974; Evans 1986; Sassen 1988). Capital has brought about a global economic scale in which space is used as a profit surface. The mobility of capital has led to changing divisions of labor globally and in the United States (Smith 1984; Sassen 1988). No country can feel it has a fixed place or destiny within the international division of labor (Lipietz 1986). Manufacturing once exclusively located in "developed" countries has relocated as multinational corporations search for lower costs and greater profits (Harrington 1984; Sassen 1988).

A major consequence of this drain of manufacturing jobs has been unemployment. Workers have been displaced from smokestack industries and it is no longer likely that large numbers of people will find relatively highly paid jobs in industry, as was once the case. The threat of job loss has been used against workers who remain to extract maximum concessions and to disempower the unions (Harrington 1984).

The general nature and dynamic of the capitalist economic system is often identified to be the source of contemporary poverty (Peet 1989; Day 1994; Gaventa 1994; Moore 1994). The structural economic forces discussed previously can be envisaged to be an integral part of the capitalistic dynamic (Wacquant and Wilson 1989). In a general sense capitalism is inherently exploitive and inequality is a natural product of this exploitation (Roby 1974; Marx 1977; Auletta 1982; Katz 1989). Capitalism has as its hallmark a contract social relationship which results in private accumulation and profit by those who own the means of production, from the labor of those who do not (Harvey 1985b; Smith 1987). The search for economic efficiency, and the neglect of social needs and justice, are part of the capitalist system. Harvey (1972) sees a lack of concern with the needs of the population, and an investment pattern based solely on the search for maximum profit. Thus, inequality and poverty are endemic and are likely to increase with time and to be accompanied by working class alienation, isolation, and self-estrangement. The emphasis within capitalism is on profit for a few and the common good is neglected (Harvey 1972; Bluestone and Harrison 1982). Underlying this neglect of the common good and drive for individual profit is the private ownership of property. Private rather than societal ownership of property results in its use for private rather than societal benefit. Those who see the critical structural problem to lie within the fundamental nature of capitalism point to the persistence of poverty as an indication that a free market lacks equalizing capacity.

Poverty for the working class within capitalism grows out of the exploitive economic relationship to encompass social and even individual psychological life (Harrington 1984; Wolff and Resnick 1987; Dear and Wolch 1989; Kodras 1992). For many scholars, other structural characteristics that may contribute to poverty are far less crucial than class divisions (Wilson 1980; 1987). Mattera (1990) sees solutions to poverty to lie in redistribution of power and wealth, control of companies and capital, income supports, a guaranteed income plan, and the creation of non-capitalist employment. Only through fundamental change in the capitalist system, and in the attendant relationship between classes, can poverty be successfully addressed. Without such basic change in the economic system improvements can only be cosmetic, tenuous, and transitory (Ropers 1991).

THE "UNDERCLASS"
THEORETICAL DEBATE

Within the current poverty literature one major perspective dominates discussions of contemporary United States poverty. This perspective centers around a group that are characterized as the "urban underclass." Theoretical debate centers on explanation of the deep persistent poverty experienced by this group. Two broad, diametrically opposed, viewpoints dominate the discussion. Each view utilizes some blend of the theoretical perspectives presented above. The explanations of poverty that emanate from this debate are often extended beyond urban poverty and are proffered as generalized explanations of poverty in the United States. Within the "underclass" debate discussions concerning any spatial dimension appear limited to intraurban dynamics related to employment.. It is the "underclass" debate that is largely responsible for the somewhat monolithic and broadly aspatial views of poverty that permeate the poverty literature. Hence, understanding of the major dimensions of the debate are necessary to understand the status of poverty research to date. Before presenting examples of these two prevalent views of contemporary poverty, a brief introduction to the "underclass" concept and debate is in order.

The urban underclass is characterized as chronically poor, but members are also often identified in individualistic terms (Hughes 1990). The underclass is said to engage in harmful behaviour and to suffer pervasive social problems (O'Regan and Wiseman 1990). Others suggest that the urban underclass is predominantly made up of racial and ethnic minorities that are undereducated and unskilled (Knox 1990). Members of the underclass are often unemployed and long-term welfare dependent (Kasarda 1990). Illegitimate births are a social ill that is frequently identified with the urban underclass (Eggers and Massey 1992), along with behaviors such as illegal drug abuse and crime which conflict with the societal mainstream (Kasarda 1990). The underclass are described by some as socially deviant, and engaged in behavior which is both self-destructive and societally destructive (Devine, Plunkett, and Wright 1992).

There is extensive and intense debate over whether the underclass exists as a bounded group (Hughes 1990; Burton 1992), whether the underclass represents a new form of poverty (Sheppard 1990), what the characteristics of the underclass are, and what underlying causes have created the phenomenon (Murray, 1984; 1988;

Wilson 1980; 1987; Eggers and Massey 1992) The label "underclass" has been criticized as implying a group apart from society and whose case is perceived as hopeless.

The Conservative-Individualistic View

A leading advocate and champion of the conservative view is Charles Murray (1984; 1988a; 1988b; 1992) who offers a view of urban poverty that lays primary responsibility on the individual and a too generous welfare system. His theory concerning the urban underclass, and poverty in general, is comprehensive and incorporates views on the role of attitudes, behaviors, government policy, reforms, and racism (Murray 1984; 1988a; 1988b; 1992).

The Charles Murray Example. The causes of the concentration of urban poverty are very clear for Murray and begin with a fundamental change of attitude which began in the decade of the 1960s. At this time the idea that not everyone who works can make a decent living began to be accepted. The poor were seen as vulnerable to structural forces beyond their control. Individual responsibility for impoverishment was replaced by the notion that society was to blame. The United States system was characterized as deeply flawed, and it was society not individuals that must change. President Johnson's anti-poverty programs, that began in the mid-1960s, were based on this shift in attitude about the root causes and responsibility for poverty (Murray 1984; 1988a; 1988b; 1992).

These social programs, Murray claims, did not eliminate poverty but rather exacerbated it. The most basic cause of poverty is individual choice, not economic structure. The plethora of social and welfare programs of the 1960s offered alternative behaviors that might be in the short term interest of individuals, but which brought long term social problems. Murray contends that the reforms of the 1960s tore apart the fabric of society, dismantling crucial norms of behaviour. It is these changes in behaviour, not structural problems within the economic system, that are at the heart of the urban underclass problem (Murray 1984; 1988a; 1988b; 1992).

Murray identifies four major dimensions within underclass poverty, all of which hinge upon change in values and behaviour rather than economic structure and opportunity. The underclass is characterized by high rates of illegitimacy, high crime rates, significant high school drop out rates, and extensive withdrawal from the labor force. While economic structural conditions actually improved conditions for the majority, including African Americans, deterioration became concentrated in the underclass as a result of their individual choices. These choices were made possible by reforms in the welfare, educational, and criminal justice systems (Murray 1984; 1988a; 1988b; 1992).

Race is not the primary issue in Murray's accounting of the racial composition of the underclass. He points to racial equality in terms of educational opportunity, occupational availability, and believes that if anything African Americans were given preferential treatment. The problem is black behaviour not racism, and he cites as evidence the economic success story of Asian-Americans. The contrast for Murray between Asian-Americans and African Americans is one of behavior. While Asian-Americans are hard working and independent, African Americans are lazy and dependent. African Americans have rejected middle class values as "white", and have been enabled by reforms put in place in the 1960s to become an underclass (Murray 1984; 1988a; 1988b; 1992).

Welfare programs have become both economically and socially enabling. Illegitimacy is no longer punished by social stigma, and now commands economic benefit rather than disincentive. A broad section of women can keep a baby without a husband and the result has been a decline in marriage. Welfare benefits also encourage unemployment. Low paying jobs have come to be viewed with scorn, and the concept that work endows self-respect has been destroyed. Social penalties for not working have been dismantled, with the result that many people, especially young black males, have withdrawn from labor force participation. Reforms in the education system have meant that schools now have an elaborate social agenda and lack previous academic and behavioural expectations. Staying in school and trying to excel scholastically are treated with derision, and school dropout is rampant. The result of this choice is that the poor lack human capital and so have limited job opportunities. Changes in the criminal justice system have resulted in a reduced chance that criminals will be caught or go to jail. Criminal activity has become a viable economic option and a means to escape from poverty (Murray 1984; 1988a; 1988b; 1992; Tootle 1989).

The reforms of the 1960s work in concert to allow a range of personal choices that result in the urban underclass. There is less pressure on both genders to marry. Males are under little pressure to support their offspring. The rewards for education cease to be of significant importance. Welfare alternatives to work exist, and criminal activity provides another economic choice (Murray 1984; 1988a; 1988b; 1992).

The policy implications that Murray identifies go beyond mere retrogression to an earlier time. The new social dynamic has taken on a life of its own and solutions need to be radical. The social and welfare programs of the 1960s that benefit the poor must be dismantled. They cannot be expected to address poverty in people who are poor because of personal choice and characteristics. The social system of penalty for deviance, and rewards for acceptable behaviour, such as hard work, striving for education, the rejection of criminal activity, and parenthood within marriage, must be forcefully reinstated. Finally, the notion of individual responsibility must be brought back. People should be forced to realize that the able-bodied must work to eat, that those without a spouse will be unable to care for children, and that those who commit crimes will go to jail (Murray 1984; 1988a; 1988b; 1992).

Governmental policy often appears to accept in some measure the conservative flawed character view of poverty. Social programs today emphasize incentives for the rich and sticks for the poor (Tobin 1990). Government, during the Reagan era, seemed inclined to accept that social programs create disincentives and in the long run make the situation worse (Gottschalk 1985; Tootle 1989).

Galloway and Vedder (1985) take an openly neo-classical view. People are poor on a voluntary basis, and make that individual choice based on personal utilities. The poor choose welfare and leisure over work, and are empowered to make this behavioral response by the ease and generosity of the welfare system. Thus, the logical course of action is to severely curtail social welfare programs (Galloway and Vedder 1985; Galloway, Vedder, and Foster 1985).

The Structural View

Placing blame for poverty on the victim has been widely criticized especially by those who emphasize the role of structure. This approach argues that individuals have little, or at least limited, control over the physical, social, economic, and political environment within which they must live their lives (Peet 1972). The problem of poverty, far from being a personal failure, emanates from external forces related to the structure of society and the economy. The various elements of, and changes in, structure previously discussed have been identified as contributing to poverty (Giarratani and Rogers 1991). Poverty is the result of broad forces (Axinn and Stern 1988), not individual action or deficiencies.

Just as Murray's view of urban poverty represents an example of a comprehensive explanation based on theory which blames individuals and government policy, urban poverty can also be explained in structural terms. Both William Julius Wilson and Michael Harrington explain the existence of the urban poor with reference to the structure of capitalism and a variety of changes within the economy (Harrington 1984; 1987; Wilson 1987).

The Wilson-Harrington Example. In contrast to Murray, Wilson identifies the urban underclass problem to be largely the result of structural economic factors that are part of the global capitalist system. The historical discrimination and migration of African Americans has made them more vulnerable to structural forces than other groups, but race is not the main issue. Nor is poverty the result of personal decisions made possible by the welfare system. At the root of the underclass problem is the shift from goods to service production, the polarization of wage levels, mechanization, the relocation of manufacturing away from the central cities, and periodic economic recession. Rising unemployment is the result of these structural forces, and in turn results in the social isolation associated with the underclass. The underclass lack access to jobs and job information networks, lack exposure to mainstream norms, and suffer from unavailability of suitable marriage partners (Harrington 1984; 1987; 1988; Wilson 1980; 1987; 1989; Davies and Herbert 1993).

African Americans form a majority of the underclass because they were disproportionally employed in industries that went into decline during economic restructuring. Also, African Americans were less geographically mobile, and thus were unable to pursue economic opportunity. Those who were able to did leave the central city, contributing to the isolation of those left behind. The tax base declined, local business shut down, and the inner city lost its leaders and role models for success. These losses have resulted in the breakdown of behavioral norms and the loss of the sense of community (Harrington 1984; 1987; Wilson 1980; 1987; 1989; Davies and Herbert 1993).

The policy implications of Wilson and Harrington's view of the causes of poverty are very different from those of Murray. Addressing the structural causes of poverty would involve reshaping the economic system. Given the deeply entrenched nature of capitalism in the United States and broad acceptance of profit as the guiding economic principle, such a revolution is unlikely. Combating the symptoms that result from the existent capitalistic system may be more realistic. One way to alleviate the impact of the economic system on the poor, lies in expansion, not elimination, of social programs. Since the basic issue is not one of personal characteristics or even of race, but rather one of class, such programs should not be racially targeted. Expanded programs should seek to address general problems of poverty and exploitation which are a result of the fundamental structure of U.S. capitalism (Harrington 1984; 1987; Wilson 1980; 1987; 1989).

SHORTCOMINGS OF CURRENT THEORY

Despite the confusing plethora of theory that falls broadly within blaming the victim or structural categories, it is evident that there remains a dissatisfaction with existing theory. For example, in a recent publication Dudenhefer (1993) finds what he considers to be the three basic theories inadequate. Neither individualistic theory, economic organization theory, nor culture of poverty theory provide a logic and complete theory of poverty (Dudenhefer 1993).

It seems clear that all of the above theories have some validity. Acceptance of the importance of a single theory in no way invalidates the possible contribution of the others. Individual choices and characteristics embodied in various forms within these theories may well contribute to poverty, as may the structural contexts in which they occur and by which they may be both circumscribed and enabled. No one theoretical perspective can hope to explain the complexities of poverty. Possible spatial and temporal variation in the relevance of the theoretical perspectives and explanations of poverty is a neglected theoretical dimension. The relative importance of the various theories may well vary both temporally and spatially. Many researchers see a particular theoretical perspective as universally primary and explanatory in the geography of poverty. Similar forces are viewed as causing poverty wherever it occurs. Mead (1992), for example, pays little attention to the geography of poverty and identifies the combination of individual psychology, the culture of poverty, and the permissive welfare state to be the ubiquitous cause. Researchers such as Smith (1991), in contrast to Mead (1992), focus directly on spatial differences in well-being. However, they identify a universal primary cause in capitalism. Space is not a backdrop for capitalism but rather is restructured by it and contributes to the system's survival. The geography of poverty is a spatial expression of the capitalist system (Lefebvre 1991; Smith 1991). However, some scholars acknowledge the importance of unique aspects of place (Johnston 1990; Johnson 1991; Kodras 1992). Both history and geography are basic interactive and recursive dimensions that articulate with such things as the capitalist system. Capitalism is not uniformly played out across space (Massey 1984), nor does it meet a homogeneous environment.

It seems likely, then, that there exists not a monolithic poverty but poverties which vary in nature (Harrington 1984). Just how these varying poverties manifest spatially is not well explored. In addition, adherence to one theoretical perspective ignores the possibility that the utility of theories may also vary depending upon place. Study of local regions which have their own unique characteristics and in which varying combinations of several universalizing theories are seen as appropriate is limited. Such studies have an important role to play in an understanding of poverty and its geographic dimensions.

III

United States Poverty and Its Definition and Measurement

BACKGROUND

Historical Poverty Levels

Poverty is a persistent problem within the United States (Anon 1989; Rubinstein 1989; Ropers 1991) and, as both Peet (1972) and Plotnick (1975) comment, absolute poverty due to insufficient nutrition still exists within the nation. Hunger, as evidenced by breadlines, is faced by some citizens everyday within the general affluence of the United States (Kodras 1992). Despite overall affluence in terms of both resources and wealth, the achievement of such affluence has persistently eluded a significant segment of American society (Burton 1992). Herbert Hoover's optimistic proclamation, made in 1928, that the day was "in sight" when poverty would be "banished in this nation" has proven to be far from reality some sixty five years later. This failure to banish poverty comes despite public awareness of the problem and the "War on Poverty" declared by President Lyndon B. Johnson in 1964.

There has been considerable progress toward eradicating absolute poverty such as is relatively common in some of the world's poorest countries. Absolute poverty is manifest in such things as severe malnutrition and lack of basic resources needed to sustain life. However, little progress has been made in addressing relative poverty. Relative poverty is defined by the values, norms, and general level of affluence of society, and people are considered poor when they fail to live at some generally accepted standard (Quibria 1991). Relative poverty is still widespread; moreover, it is only the growth of transfer payments that has prevented an increase (Smolensky 1981-82; Greenstein 1985). Since the political leadership in the 1980s with its ideological opposition to welfare curtailed many transfer payments, poverty is once more affecting increasing numbers of Americans (Dobelstein 1987; Tootle 1989). Moreover, it can be convincingly argued that many who live above the official poverty level experience extreme deprivation; thus the number of poor Americans may be even larger than official statistics suggest.

During the post-industrial history of the United States the extent and nature of the poverty problem has undergone broad changes. The widespread absolute poverty that existed across the United States worsened and deepened during the years of the Great Depression; in these years poverty was a "normal" condition (Harrington 1987). The 1930s were an important era that resulted in changes in policy toward poverty and attendant changes in the extent and depth of deprivation. Prior to the depression years such public relief that did exist was based on English poor laws, was punitive in nature, and was administered locally. Poverty in the 1930s became a huge problem, and the threat of civil unrest forced the federal government to intervene (Kodras 1992). A host of federal programs were instituted and thus after the Depression there was a long-term decline in poverty (Ross 1987). While generally in long-term decline, within the post World War II period some shorter-term trends may be noted.

In the period of national growth that followed World War II, the number of people in poverty declined (Anon 1989). Between the years 1929 to 1960 poverty was reduced at an annual rate of 2.2 percent, but this decline was not consistent over the post-war years. From 1935 to 1947 the decline in poverty was rapid (4.8 percent per year), as the United States completed the transition from an agricultural to a mass production oriented economy. The decline in poverty slowed considerably in the early 1950s; from 1953 to 1960 the rate dropped to 1.1 percent (Conference on Economic Progress 1962). At the end of the decade of the 1950s one fifth of the nation remained in poverty and two fifths lived in deprivation, when deprivation is taken to mean lack of a modestly comfortable level of living. Nor did mere poverty rates tell the whole story. In 1960 over 12.5 million Americans had incomes less than half the official poverty line, a condition that must be envisaged as abject (Conference on Economic Progress 1962).

Lyndon B. Johnson's State of the Union Address was delivered in January 1964 when progress against economic and social deprivation had lost its momentum. This address contained policy statements that were to give much needed impetus to the decline in poverty that had occurred since the end of the Depression. President Johnson's "unconditional declaration of war on poverty" brought sweeping changes that were to impact American society over the next twenty five years. The result was that general poverty levels continued to decline until 1973 (O'Hare 1985). Within two years of Johnson's declaration of war, his rhetoric had been transformed into social programs and the liberal era of federal policy had begun (Haveman 1987).

In 1964, nineteen percent of Americans had incomes below the official poverty line. By 1973, poverty rates had reached a record low of almost 11 percent (O'Hare 1985) (Figure 1). In the years between 1962 and 1973 more Americans began to enjoy a minimum living standard as the number of poor people fell from thirty-eight to twenty-three million (Figure 2). Despite more people attaining a minimum living standard and escaping from absolute poverty, relative poverty remained virtually unchanged, a situation which reflected continued inequality (Plotnick and Skidmore 1975).

After the record low levels of poverty experienced in 1973, poverty levels from 1973 to 1979 remained relatively constant (O'Hare 1985). During these years between 11 and 12 percent of Americans lived below the official poverty line (Figure 1). 1980 was a turning point with regards to poverty trends within the United States. Poverty began to rise at a time when the nation appeared to be becoming increasingly affluent (Eberstadt 1988). The deterioration was rapid. In 1979, 11.7 percent or some 26 million Americans were officially categorized as poor. Just four years later, in 1983, the number of poor Americans was over 35 million or 15.2 percent of the population. This 1983 poverty level was the highest level since Johnson declared his war almost twenty years earlier (O'Hare 1985). Kodras (1992) attributes this dramatic reversal to change on two fronts. First, both the economy and nature of industrial production have undergone a dramatic transformation. Since the 1970s the United States has been losing its dominant position in the global economy, while at the same time industry has moved away from mass production to more flexible systems. The second change has been in the realm of social programs. Welfare is poorly designed to address new forms of deprivation. In addition, during the early 1980s and the dawn of Reaganomics, welfare began to be perceived negatively and social spending cut back (Axinn and Stern 1988; Kodras 1992). During the 1980s a war was waged on welfare (Axinn and Stern 1988). Aid to Families with Dependent

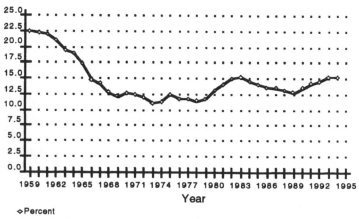

Figure 1. Percentage of People in Poverty - 1959 to 1994

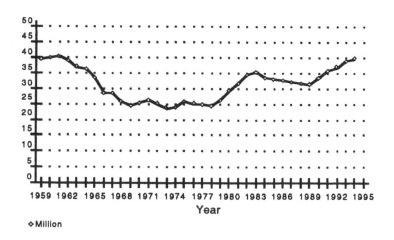

Figure 2. Millions of People in Poverty - 1959 to 1994

Children (AFDC) and Medicaid were effectively cut back as neither the income limits or benefits kept pace with inflation. Other more direct cuts were made to fulfill requirements such as those contained within the Gramm-Rudman-Hollins targets (Greenstein 1986). The Reagan administration not only cut welfare, but also adjusted the tax system to become more unequal. Taxes for rich Americans were cut and tax shelters expanded (Ropers 1991). During this time the depth, as well as the extent of poverty, increased as those already poor were impacted by cuts in social spending. By 1983 a widening poverty gap indicated that the poor were becoming poorer (O'Hare 1985). There had been a polarization of wealth in the United States as the gap between the richest and the poorest Americans widened (Goldstein 1986; Kahan and McKeown 1986; Ropers 1991). In recent years, the polarization of wealth in America has been increasing (Ropers 1991). In 1980 the poorest fifth of American families commanded 5.2% of total income and this had fallen to 4.5% by 1991. In contrast, the most affluent one fifth of families enjoyed 41.5% of income in 1980 and their share had increased to 44.2% by 1991 (U.S. Department of Commerce 1992a). This increasing inequality not only has consequences such as homelessness for individuals, but also has an impact on broad economic and social structure (Ropers 1991; Anglin and Holcomb 1992).

After the poverty rate maximum in 1983 the remainder of the 1980s saw a slight decrease in levels of poverty (Eberstadt 1988), which was perhaps a cause for optimism. The downward trend in poverty levels, although slow due to wage stagnation and Reagan policies (Solow 1990), seemed consistent. However, as the 1990s began poverty levels began to rise once more. In recent years the federal government has increasingly begun to abdicate responsibility for social programs that address the poverty problem. These responsibilities are being shifted to state and local government as well as to the private sector (Plotnick 1990; Kodras 1992).

Contemporary Poverty

General poverty levels in the 1990s appear to be on the increase, and the long term decline that occurred in earlier years has not continued (Tobin 1990). By 1994, 15.3 percent of the U.S. population had incomes below the official poverty line (Figure 1). In terms of absolute numbers of the poor the 39.7 million Americans impoverished in 1994 constitute the largest group of poor since 1961, prior to the war declared by Johnson (U.S. Department of Commerce 1992a) and extensive anti-poverty programs began to be conceived and enacted (Figure 2). Poverty levels within the United States continue to be severe when compared to other industrialized countries (Hoppe 1991). Using official definitions of poverty, it is clear that poverty has increased within the United States during the early 1990s. Pollard (1992) offers some statistical evidence drawn from the March 1992 issue of the Current Population Survey published by the Bureau of the Census.

Both average household real incomes and Per Capita Incomes (PCI) fell during the first year of the decade of the 1990s. The drop was substantial; incomes dropped by 3.5 percent, while PCI declined by 2.5 percent. In conjunction with this decline in incomes came an increase in those officially classified as poor. This increase was in terms of both the absolute numbers of individuals and families, and the percentage poor. By 1991 2.1 million more individuals within 600,000 families had fallen into poverty since the previous year. The poverty rate had increased from 13.5 to 14.2 percent (Pollard 1992).

Even when one goes beyond official counts of poverty it is clear that poverty has increased in the decade of the 1990s. Taking into account taxes paid and non-cash benefits received makes little difference to conclusions concerning general poverty trends. While such a definition yields a lower percentage deemed to be in poverty in 1991 (11.4), the trend was still upwards in terms of poverty levels (Pollard 1992). Some alternative measures of poverty yield an even grimmer picture than do official statistics.

Ruggles (1992) calculated poverty levels for 1988 using three alternative, and perhaps more realistic, poverty measures. Using an Engel's coefficient1 of 6 rather than 3 almost doubles the poverty rate from 13.0 to 25.8 percent. If one defines the poor as those who command incomes at 50 percent or less of median income, the poverty rate increases to 19.5 percent. The third alternative measure which calculates the poverty threshold based on housing consumption results in 23 percent of Americans being classified as poor (Ruggles 1992).

The poverty levels and changes experienced have differed among various groups within the population. Female headed households fared less well than other groups. The real income of female headed families dropped by 5.4 percent from 1990 to 1991, while the percentage of such families in poverty rose from 33.4 to 35.6 percent. It is often children that make up the majority of such families, and by 1991 it was children who were the most likely of any age group to be poor. In 1991 21.8 percent of all children were poor, and children constituted over 40 percent of the poverty population (Pollard 1992).

Changes in poverty were experienced differently by various racial and ethnic groups. Asian Americans enjoyed the highest median income in 1991 of any group, but also experienced the biggest decline in income. Despite a decline in income, the absolute number and percentage of Asian Americans who fell below the official poverty level remained stable (approximately 13.8%) between 1990 and 1991. Poverty rates for African Americans remained high (32.7%), but unchanged from 1990-1991. The median income for African Americans, the lowest for any group, also remained unchanged. Poverty has long been endemic in the African American population, and the 1990s have brought little relief to this subsection of the American population. Similarly, Hispanic Americans continued to have a low median income only slightly above that of African Americans. Despite the relative stability of Hispanic median income, the number of poor Hispanics increased during the early 1990s. The fortunes of white Americans in the year from 1990-1991 grew worse. The median real income for whites fell, and some 1.4 million more white Americans fell below the poverty line (Pollard 1992).

Poverty continues unabated in America's urban centers, and this poverty commands much public and academic attention. However, poverty in 1990 was not confined to urban areas but also continues to persist among the rural population, often at a rate that rivals that found in the inner city. Dudenhefer (1993) points out that the rural poor in 1990 are more likely than the urban poor to be chronically impoverished. They are less dependent on welfare and more likely to be working, and in addition are less likely to be single parents than are their counterparts in the city. The rural poor more often live in married couple families than the urban poor. Thus, it seems the rural poor exist in poverty despite struggling to remain part of the economic and social mainstream (Dudenhefer 1993).

Poor Americans in 1990 are seen to be distributed spatially in similar fashion as they were in 1980. It is the central cities that have the highest poverty rate within urban areas, although poverty in the suburbs is by no means unknown. Poverty rates in the central city are approximately 19 percent compared to 8.7 in the suburbs. The rural poor continue to be found predominantly in the South. Approximately 55 percent of the non-metropolitan poor live in this region. Little seems to have changed since 1980 in terms of poverty, except perhaps that it has become more widespread (Dudenhefer 1993).

Devine et al. (1992) contend that contemporary poverty has a different character than in the past. They conclude from research that impoverishment today is more destructive and intractable than it has ever been, as U.S. society becomes increasingly unequal and polarized (Devine, Plunkett, and Wright 1992). Poverty, it seems, has not been eradicated by the "war" declared in the 1960s, nor does it seem credible that progress will be swift and easy as some would contend (Morgan et al. 1962). Instead it appears likely that poverty will continue to be one of the major problems facing the United States as we enter the twenty first century.

The above discussion provides an overview of poverty as it existed and continues to exist in the United States. Such a discussion is useful as a preliminary and superficial understanding of the problem. However, this general view does not address some important issues concerning the difficulties of both defining and measuring poverty. A deeper understanding of poverty and the limitations of such general discussions requires that issues of definition and measurement be explored.

DEFINITION AND MEASUREMENT
OF POVERTY

Introduction

Whether perceived as problematic or not, it is apparent that poverty has been recognized as a social condition for much of the history of mankind. In such ancient works as the Christian Bible poverty is referenced as long existent. History has recognized poverty and yet has reached no consensus as to its definition or measurement (Burton 1992). No one standard exists either spatially or temporally (Valentine 1968). This lack of consensus is readily apparent in contemporary writings in which scholars disagree with regard to the extent and tractability of poverty (Bassett 1973; Wohlenberg 1970). Morgan et al. (1962) state that poverty in the United States has recently shown a marked decline and could "be abolished easily and simply by a stroke of the pen." Larner and Howe (1970) present a diametrically opposed view by stating, "The war on poverty in America has barely begun... eradication of poverty is not a simple matter, to be effected through one piece of action or slogan... ." Varying definitions of poverty and methodology of measurement are potentially major factors in this disagreement concerning poverty, a disagreement which is often attributed almost entirely to ideological differences.

What poverty is, and how it can be measured, is of interest to many, including academics, politicians, and social practitioners. It is a crucial and fundamental question that must be addressed by any who embark upon research related to inequity (Atkinson 1987). Before investigating the spatial distribution of the poor, delineating poverty regions, identifying the characteristics of the poor, and discussing possible causes of poverty it is vital to discuss the definition and measurement of the phenomenon under scrutiny.

Historical Perspectives

Historically, definitions of poverty have been based on either economic or sociocultural criteria (Oster, Lake, and Oksman 1978). Prior to the twentieth century, definitions were purely economic and largely ignored the sociocultural characteristics of the poor. Definitions in favor within the United States were extensively influenced by British precedents (Katz 1983; Quibria 1991). Prior to industrialization and expansion of the economy beyond simple agrarianism, the subsistence level constituted the line which separated the poor from the non-poor. The general level of affluence and economic base could permit no other standard. The subsistence level may be described as the minimum food, clothing, and shelter required to sustain life (Oster, Lake, and Oksman 1978). Despite the seeming simplicity and clarity of this definition, even the meaning of subsistence is open to interpretation. Just what constitutes sustaining life, and what the basic minimum physiological requirements for such a life are, may be open to debate (Goldstein and Sachs 1983). Perhaps in the case of food a modicum of agreement may be observed on what the level of subsistence is; however, just what level of health and capacity for work is to be supported in not clear. The subsistence level in terms of clothing and shelter is even less well established (Rein 1969; Batchelder 1966). Even in contemporary times the definition of subsistence varies with culture. Setting a poverty level requires fundamental social judgments that are linked to social norms. There is no right way to measure poverty, and no measure is value free or free of moral implication (Ruggles 1990; 1992; Haveman 1992/1993).

Economic definitions up to the late nineteenth century were developed and administered by local communities, often in very arbitrary ways. These definitions were underpinned by value judgements concerning the characteristics of those who were impoverished. The general view of the poor was that they were weak, immoral, idle, purveyors of their own condition. There was little sympathy for those suffering from poverty, and no interest in development of consistent poverty definitions and ways of measurement.

The late nineteenth century brought increased interest in science and objectivity. As a part of this broader interest in science and the rejection of things arbitrary, came attempts to rationalize definitions of poverty. The first definition of a poverty line was made in Great Britain by Booth in the 1880s (Goldstein and Sachs 1983; Quibria 1991), and much research on similar poverty lines quickly followed in the United States (Appelbaum 1977). An income level which constituted a poverty line was tied to basic needs and subsistence. In the late nineteenth century Booth and Rowntree were the first to to develop a hypothetical market basket that would supply basic needs (Rowntree and Lavers 1951; Zimbalist 1964; Booth 1970). This basket was based on the nutritional requirements of convicts for subsistence, as identified by Atwater (Oster, Lake, and Oksman 1978). Thus, steps were taken toward standardization of the definition and measurement of poverty. Despite these advances poverty remained essentially defined by subsistence.

The twentieth century brought increasing affluence to the newly industrialized nation of the United States. The economy rapidly expanded and the general populace became consumers of a whole host of goods. With this affluence and consumerism came a change in attitude concerning definitions of poverty. Definitions of poverty began to have a relative dimension. Poverty came to be viewed as relative to the general standard of living, and instead of subsistence such phrases as decent standard of living began to be used. In Western societies, including the United States, perceptions of deprivation expanded beyond the mere lack of ability to survive (Oster, Lake, and Oksman 1978).

Changing perceptions concerning the definition of poverty resulted in national government intervention. The Great Depression and Roosevelt's "New Deal" in the 1930s catapulted the federal government off the sidelines and out of its role of publicist of issues with which the states must grapple. National government became involved with a multitude of programs for the poor (Katz 1983; Dudley 1988). It is in this era of federal government involvement that the roots of contemporary poverty definitions lie.

Current Perspectives

The Official Poverty Line. Just as throughout most of history, economic definitions of poverty are currently much in evidence. The measure almost universally used by government is an income poverty line. This poverty line was developed in a series of stages beginning in the 1940s (Hoppe 1991).

In 1941, an important milestone was the development of Recommended Dietary Allowances that were later used as a basis for United States Department of Agriculture (USDA) food plans (Oster, Lake, and Oksman 1978). A series of food plans became a cornerstone of modern need definition.

In the 1960s President Lyndon B. Johnson and the United States Congress declared a war on poverty with the passage of the Economic Opportunity Act (Smolensky 1981-82; Arkansas Department of Local Services 1977). As part of the war on poverty came the first systematic study of poverty (Quibria 1991), and a market basket of goods deemed to fill minimum needs, as defined by USDA food plans, was developed. With 1961 as the base year, this market basket was linked to the Consumer Price Index in subsequent years (Oster, Lake, Oksman 1978). The USDA originally developed food plans at three different economic levels, the Low Cost, the Moderate Cost, and the Liberal. From these food plans a market basket could be derived at differing dietary levels (Oster, Lake, and Oksman 1978). An income level, or threshold, that could be used to divide the poor from the nonpoor could then be defined by applying the Engel's coefficient to the cost of the designated food basket (Rein 1969; Ruggles 1992). The Engel's coefficient is the reciprocal of the fraction of income that must be spent on food. The official poverty levels used an Engel's coefficient of three; this estimate that one third of income was spent on food came from a food consumption survey conducted in 1955 (Ruggles 1990; 1992). The selected coefficient was applied to the Low Cost food budget that was considered to provide for minimum dietary needs.

A problem was apparent with this developed poverty level. There were too many people in poverty for political comfort. A purely definitional adjustment was quickly made. Thus, in 1964 the Economy Food Plan was substituted (Goldstein and Sachs 1983) at a level of only 75 to 80 percent of the original Low Cost Plan. From the outset the Economy Plan was never intended to provide an adequate permanent diet. Instead it was intended only for temporary or emergency use when funds were low (Orshansky 1969). Despite its admitted inadequacy the

Economy Food Plan remained the standard used for the definition of poverty until 1974. In July, 1974, the Thrifty Food Plan replaced the Economy Plan (Goldstein and Sachs 1983) establishing an even more inadequate dietary minimum than the temporary or emergency Economy Plan (Appelbaum 1977). Thus, reported changes in the percentage and absolute numbers of the poor reported by the federal government must be viewed with the knowledge that changes occurred in the definition of poverty. By switching food plans which form the basis for the definition of the official line, some of the poor suddenly were redefined as nonpoor with no change in their real world circumstance.

Prior to the early 1960s poverty level establishment took no account of family size. The line was set at the same level for a family of three as a family of ten or even more. This resulted in mis-categorization of some large families as non-poor (Conference on Economic Progress 1962). In 1963-64, Mollie Orshansky of the Social Security Administration conducted a series of studies in an attempt to formulate a more flexible system of poverty levels. The Orshansky system set poverty levels that varied with family size and type (Bassett 1973; Ruggles 1990; 1992). The Orshansky system was based on minimum nutritional need as established by the Economy USDA food plan (Ferman, Kornbluh, and Haber 1969; Bassett 1973). The system considered money income (Goldstein and Sachs 1983), but also placed families into 124 groups based on family size, sex of head of household, age, and farm/non-farm residence.

Attempts have been made to set other, often more generous, poverty lines. Notably those lines developed by the economist L.H. Keyserling (Ferman, Kornbluh, and Haber 1969), the President's Council of Economic Advisors (Ferman, Kornbluh, and Haber 1969), and The Bureau of Labor Statistics (Vitelli 1968; MacLachlan 1974). However, since 1969 it is essentially the Orshansky system that has been mandated for use for most governmental and reporting purposes (Ruggles 1992).

Despite the extensive and continued use of the Orshansky system, criticisms have been leveled against this system on various grounds. Some criticisms have been addressed and the system adjusted; some have not. Until relatively recently it was assumed that farm residents received some food and housing benefits directly that supplemented cash income. Rural families were judged to need forty percent less cash income than non-farm residents (Orshansky 1969). Goldstein and Sachs (1983) saw little evidence of the validity of this assumption, and Oster et al. (1978) suggested a rural/non-rural distinction would be more realistic and appropriate. The federal government acknowledged the validity of this argument against different poverty income standards for farm and non-farm residents, and ceased to distinguish between families on that basis in the early 1980s (Hoppe 1992).

A major criticism of the Orshansky system is that it is unable to adjust to changing family needs and patterns of consumption. As Ruggles (1992) aptly observes, in 1955 families had little need of such things as telephones and child care, services that are essential to many families today. The market basket used reflects consumption in the 1950s, and remains unresponsive to general changes in family needs. Even when first formulated, the contents of the basket were based on the buying patterns of the average consumer, not the low income shopper (Oster et al. 1978).

Even the adjustments made for changing prices may be unrealistic. These annual adjustments to the cost of the market basket are linked to the total Consumer Price Index, whereas the necessities on which the poor must spend a greater percentage of their income than the more affluent, such as food, housing, and transportation, tend to increase at a differentially higher rate than the index overall (MacLachlan 1974). Housing, in particular, has consumed increasingly high percentages of the average family budget. In 1955 an average family spent approximately 33 percent of its income on housing; by 1992 this percentage had risen to about 42 percent (Ruggles 1992).

Another criticism is that the market basket approach assumes expert and optimal buying skills and habits on the part of low income consumers (Oster, Lake, and Oksman 1978). The reality is that low income shoppers are likely to be the least educated of consumers, and may well lack the information and skills to use their income to maximum effect. In addition, bulk buying to minimize cost is often very difficult for the poor who lack the needed capital and storage facilities such as a freezer. A final problem for the poor is that they often have few shopping choices and provide a captive market for food suppliers. Vendors may charge high prices for the goods they offer, which the poor have little option but to pay.

Many have argued that the poverty lines identified by the Orshansky system are unrealistically low. Even Orshansky acknowledges the standard to be conservative (Orshansky 1969). The Economy food plan used as a basis for poverty lines was questionable as a standard of adequacy for a permanent diet. Further, the Engel's coefficient of three that was applied in order to arrive at a monetary standard was probably an inadequate multiplier even when first applied in 1963. Many, more or less conservative estimates of the appropriate Engel's coefficient have been put forward but most seem to agree that 3 is too low. Goldstein and Sachs (1983) submit that a factor of 3.4 was a more accurate multiplication factor by 1983 (Goldstein and Sachs 1983). Ruggles (1992) reports that consumer data indicated that as early as 1960-61 food expenditures accounted for approximately one quarter rather than one third of a family's budget (Ruggles 1992). It is likely that the Engel's Coefficient of three which is used no longer represents that portion of a low income budget that must be spent on food. Ruggles (1992) identifies a realistic coefficient for the 1990s to be six. Thus, the 1991 poverty line income of $13,924 for a family of four is thought by many to be hopelessly inadequate.

The problem of using a food plan less generous than one considered for emergency or temporary use, as a basis for defining poverty lines, has already been stated. Inherent within the use of any of the USDA food plans is the difficulty that nutritional needs vary not only with age and sex, but also with the level of physical activity and individual physiology (Rein 1969). It is difficult to define minimum nutritional needs (Quibria 1991).

Another major problem within the official definitions of poverty, of particular interest to geographers, is that no adjustment is made for regional variations in cost of living (Goldstein and Sachs 1983). Nor is the availability of government programs taken into consideration (Oster et al. 1978). As Oster et al. (1978) indicate, it is difficult to assess how much lower costs of living offset lower incomes in regions such as the South, but no provision is made within poverty levels for even large regional variations (Cowell 1977; MacDonald 1969).

Those who believe that poverty should be defined relative to the general affluence of society identify another problem with Orshansky's system of poverty definition. While adjustments are made for changing costs, no adjustment is made for changing general levels of affluence. The result of this deficiency has been that poverty lines have declined with respect to median income. In 1967 the poverty line for a family of four represented 43 percent of median family income; by 1988 this percentage had fallen to 37. Thus, the official definition of poverty lacks relativity to changing general levels of income. While poverty levels may appear unchanged, inequality may be substantially increasing (Ruggles 1992).

Another problem with the income identified to constitute poverty is that it is pre-tax income that is considered (Pollard 1992). This seems somewhat nonsensical since money paid in taxes is not available to purchase needed goods and services (Ruggles 1992).

Finally, poverty may be overestimated in that assets are not included when assessing a family's level of affluence (Oster, Lake, and Oksman 1978), and in-kind income such as food stamps are also not included (Goldstein and Sachs 1983; Ruggles 1990; 1992). As Vitelli (1968) indicates, the old who have accumulated assets over a lifetime may need less income to maintain a specific standard of living than a young family which has no such assets.

Alternatives to the Official Poverty Line. One alternative for measuring and defining poverty involves continued use of the market basket approach. Since the main criticisms of this approach are insensitivity to changing patterns of consumption and levels of general income, suggestions have been made that would address these problems. Ruggles (1992) suggests re-estimation of the market basket of needs at regular intervals (Ruggles 1992).

When using income as the criteria for poverty definition, other methods besides the link to food adequacy can be utilized that are more relative to society's general level of affluence. Simplistically, one could define a lowest quantile of the income distribution as living in poverty. While this might be useful in an investigation of the characteristics of people with relatively low income, obviously no changes in the extent or depth of poverty could be determined as both would remain statistically invariant (Oster, Lake, and Oksman 1978). Such a measure would seem to be somewhat meaningless, since the actual conditions experienced by the lowest income group are dependent on both the general level of affluence as well as on the nature of the income distribution. In an egalitarian society the lowest income segment of society would experience similar living conditions as the most affluent group. The living standards of both could be enriched or impoverished in absolute terms depending on general levels of development.

Another possibility is to define poverty in relation to median income. Such definitions are usually expressed as a percentage of the median (Goldstein and Sachs 1983). While using the median income as a reference does allow for comparison in extent and depth of poverty over time, an arbitrary decision must be made as to what percentage of the median constitutes poverty (Oster, Lake, and Oksman 1978; Fuchs 1992). Varying percentages have been used within government circles. For example a 38 percent level was in use in 1979 and a 54 percent level in 1960 (Harrington 1984), and definitions as high as 60 percent have been commonly used (Goldstein and Sachs 1983). Ruggles (1992) makes the point that if such a poverty standard is used, the numbers of those statistically defined as poor is likely to rise most rapidly during periods of economic growth. The real incomes of the poor may well be rising during such periods, while at the same time poverty is deemed to be increasing (Ruggles 1992). It is clear that such relative poverty definitions can create conceptual anomalies. Under such definitions only a change in the shape of the income distribution as a whole can reduce "poverty" (Fuchs 1992; Ruggles 1990; 1992).

Recent research has focused on the refinement of income measures of poverty (Renwick and Bergman 1993; Rodgers and Rodgers 1993). Several different poverty indices have emerged from the econometric literature which seek to improve upon a traditional simple head count of persons who fall below some designated income level. Rodgers (1991) compared several of these more complex indices with the traditional head-count measure. He concluded that these indices correlated strongly with the simple head-count and also were highly intercorrelated. It seems, therefore, that attempts to develop sophisticated income measures have led to minimal improvement (Rodgers 1991).

While a majority of definitions and measures of poverty focus on income, not all do so (Fuchs 1992; Delhausse et al 1992). Although it is apparent that the current poverty line is set in terms of food subsistence level and monetary income, government programs aimed at alleviating poverty seem to reflect a broader view (Rein 1969). Programs are aimed not only at improving income, but also at providing health care, decent housing, educational opportunity, and a wide range of community services. As Senator Ribicoff observed in 1972, the 170 anti-poverty programs had an annual cost three times the cash needed to bring all the poor up to the poverty line (Goldstein and Sachs 1983).

It seems that the range of government programs move beyond the official definition of poverty toward expanded definitions such as expressed by Harrington (1984). Harrington views as poor individuals and families whose resources are so small that they are excluded from modern life, which consists of the common patterns and activities of the societies in which they live. Harrington's works (Harrington 1963; 1984) in particular have spurred interest in non-income measures and concomitants of poverty. In the view of those who advocate broad measures, economic poverty is merely a symptom of a broader inequality, entrenched by a lack of economic, social, political, and personal power (Turner and Starnes 1976).

The nonincome measures of poverty are numerous. They may include assets, such as a house or car (Cowell 1977), employee benefits, and legal services (Harrington 1984). Also included are consumption of and access to specific items such as telephones, hot water, and gas (Vitelli 1968), access to and quality of community services and facilities such as parks and libraries, as well as in-kind benefits such as food stamps and medical care (Cowell 1977; Goldstein and Sachs 1983). Other measures are the possession and availability of education and training (Cowell 1977), and opportunity for employment (Levitan and Shapiro 1987). Whole communities can be viewed as impoverished when using such broad definitions. This can be argued where resources such as worker skills and low wage levels lead to lack of community facilities and inferior quality of services (Goldstein and Sachs 1983; Ferman, Kornbluh, and Haber 1969).

The problems of valuing and measuring non-income elements of impoverishments are manifest and multifarious (University of Wisconsin-Madison 1981-82; U.S. Bureau of the Census 1982; Anon 1989). Poverty lines based on such non-income definitions are more ambiguous and difficult to draw (Rein 1969). How does one value in-kind income, or access to such things as legal services and medical care that do not provide an improved life style per se (Goldstein and Sachs 1983)?

Some researchers suggest the use of consumption as a measure of poverty. Use of consumption, it is argued, provides a better picture of the quality of life experienced than does any income measure. Lack of consumption, and patterns of consumption substantially different from social norms are said to indicate deprivation relative to society (Delhausse et al.1993; Slesnick 1993). However, such consumption based measures appear to ignore the influence of personal preference, and cultural preferences that may vary spatially.

Ruggles (1992) suggests using housing costs to define a poverty line. In much the same way as basic food costs are used as a basis for drawing the official line, it is possible to use housing costs in

similar fashion. Using housing costs to derive a poverty standard offers
an alternative, but the result is a unidimensional income definition of
poverty. Moreover, standards of adequate housing vary much as do
standards of adequate diet (Appelbaum 1977).

Inclusion of nonincome criteria into poverty definition and
measurement adds a seemingly desirable and humanistic element.
However, no practical way to operationalize such definition and
measurement has been proposed. Discussion of the non-income
elements of poverty serves to accentuate the complexity of the issue
(Ferman, Kornbluh, and Haber 1969). The inclusion of non-income
poverty measures gives a profile of individual cases rather than
delineating any cohesive group as identifiably impoverished (Oster,
Lake, and Oksman 1978)

In the decade of the 1970s, definitions of poverty were further
refined and complicated by the increased interest in socio-cultural factors
(Oster, Lake, and Oksman 1978). Behaviour and attitudes were seen as
important aspects of poverty that in large measure were responsible for
transmitting poverty across generations (Ferman, Kornbluh, and Haber
1969). "Culture of poverty" began to be a popular phase and was seen
to be characterized principally by attitudes of defeat and powerlessness
in conjunction with an enveloping pessimism (Oster, Lake, and
Oksman 1983; Ferman, Kornbluh, and Haber 1969). Such attitudinal
and behavioural characteristics of the poor, while probably initiated by
economic deprivation, may not correlate with lack of income or wealth
in the long term. Rather the culture of poverty may be self perpetuating
and require more than economic input to break the cycle (Oster, Lake,
and Oksman 1983). As Ferman et al. (1969) also point out, it is likely
that no one culture of poverty exists, but that distinct cultures of
poverty can be observed divided along ethnic and geographical lines. In
any event, it would be clearly impractical, as well as controversial, to
define and measure poverty using such behavioral, attitudinal, and
cultural criteria.

The infusion of yet more parameters into a definition of
poverty serves to capture the reality of the condition more completely
and more poignantly. However, what should be clear is that as the
criteria used become broader and more sophisticated the definition of
just what poverty is becomes more nebulous. At the same time,
measurement becomes more difficult and impractical. It is one thing to
conceive of a global view of poverty but it is quite another to measure
it, and such global measures lack usefulness in temporal or spatial

studies. Batchelder's (1966) point is well made that what is often needed is a "quick and dirty" definition that cleanly distinguishes between the poor and non-poor , and which can be readily calculated with available data.

Discussion

Definitions and measures of poverty vary from absolute to relative and from purely economic to global (Ropers 1991). Direct comparison of incomes in constant dollars through time denigrates the relative dimension of poverty (Conference on Economic Progress 1962). That poverty must be viewed relatively can be proven apagogically: use of a fixed standard through time and space leads to the conclusion that there are no poor in the United States today (Ornati 1969).

Although purely economic measures of poverty have been widely criticized, noneconomic alternatives have proved difficult to operationalize. Discussions of poverty have become global, but the condition's definition and measurement remain almost invariably tied to income. The most frequently used income measures are income with respect to median income, official poverty lines, and Per Capita Income (PCI). Incomes used relative to medians do accommodate changing general levels of affluence, but poverty levels thus derived vary with the perceptions and politics of the time. As has been discussed, it is an arbitrary and value laden decision just what percentage of the median constitutes poverty.

The use and problems of the official poverty lines have been discussed. Despite these well known problems with the official poverty line, however, poverty thresholds defined by money income remain the widely used official poverty measure.

Within the academic world PCI is commonly used (Bassett 1973), in conjunction with some standard, to define and measure poverty. The use of PCI has both advantages and disadvantages. Data on PCI are relatively reliable and readily available (Wohlenberg 1970), but the level of PCI judged to constitute poverty has not remained fixed. Another consideration is that factors such as high child mortality rates may raise apparent PCI, but few would argue that this represents less impoverishment.

Definition and measurement of poverty also depend on goals and intended use. Complex definitions of poverty that include income, wealth, opportunities, resources, skills, education, health, behaviour, and attitudes, provide a view of the realities and complexities of an impoverished existence. As such they are useful in studies aimed at identifying who the poor are, and what the extent and nature of poverty is. For use within government programs and for viewing changes in poverty through time they have limited usefulness. Utilizing the whole gamut of poverty characteristics is an attractive vision, but who would agree on what these characteristics are or on their degree of significance? Even the "poor" themselves do not always agree with this assessment of their condition by those of us who seek to formulate a definition of poverty (Ferman, Kornbluh, and Haber 1969).

In practical terms many studies must accept very inadequate, simplistic proxies of true poverty for use in definition and measurement. Constraints of administration and of available data make the use of such naive ideas inevitable. Setting poverty lines based on income is relatively simple, convenient, and practical. However, no equivalent poverty lines of any kind exist through time (Ferman, Kornbluh, and Haber 1969).

Despite lack of agreement on a definition of poverty, some broad characteristics do seem identifiable (Peet 1972) beyond merely low income. While not always practical for use in defining and measuring poverty, these dimensions of impoverishment are important in discussion and analysis. Poor living conditions are highly correlated with low income (Smith 1972). Housing quality has been the target of anti-poverty programs such as the Housing Assistance Program, which was administered at a cost of $5 billion in 1980 (Haveman 1987). Characteristics of education such as achievement, duration, and quality, are also conditions associated with income poverty (Smith 1972). Education improvement programs such as Head Start and Upward Bound show government to be cognizant of the link between poor education and income poverty (Haveman 1987). Poor access to, and quality of, health care is also associated with income poverty (Fuchs 1992). While poor housing and lack of education may be a result of personal choice, good health is almost universally desired. Thus, it may be in poor health that poverty manifests itself in the most fundamental way. Programs such as food-stamps and Medicaid attempt intervention in the health facet of poverty (Haveman 1987). In a market economy such as the United States accepting some income level as a proxy for broader poverty may not be problematic. However, other aspects of poverty must not be entirely dismissed for the purpose of convenience. The inclusion in a study of poverty of variables within housing, education, and health may result in a better measure of material poverty and deprivation than income statistics alone can provide (Eberstadt 1988).

IV

The Nature of Poverty and
Characteristics of the Poor

INTRODUCTION

Poverty, as has been expressed earlier, is more than the simple lack of income (Mingione 1991). Impoverishment is multifaceted, and numerous dimensions have been identified by various researchers. Some widely accepted dimensions of poverty that lie beyond lack of income are in the areas of health, education, and housing (Smith 1972; Holman 1978; Smith 1979; Bray 1995; Scheller 1995). In addition, poverty does not affect all groups equally. Certain sub-populations, for example children and African Americans, are at high risk of being poor (O'Hare 1985). Finally, the poor are not evenly geographically dispersed, and, thus, poverty has a spatial dimension (Brunn and Wheeler 1971; Morrill and Wohlenberg 1971). In order to gain a more detailed and realistic view some major facets of poverty, the broad characteristics of the poverty population, and general spatial distribution of the poor, must be discussed.

DIMENSIONS OF POVERTY

Health

A major dimension of poverty in America that is related to, but not circumscribed by, income centers upon health (Fuchs 1992). In the mid 1980s, approximately 34 million people had no health care insurance (Greenstein 1986). Lack of health care, poor nutrition, onerous working conditions, substandard housing, and a generally poor environment all contribute to the health problems faced by those with low income. Poor health and low income exist in a vicious cycle. Low

53

income denies access to health care, good nutrition, and decent living conditions. Low income jobs often involve physically demanding work in poor conditions (McCormick 1988). Poor health reduces the capacity to work and limits individual activities so that few opportunities exist to raise income.

A major manifestation of the health dimension of poverty is high rates of infant mortality. Infant mortality is a strong indicator of social conditions and life chances (Smith 1979; Mangold, Schwab, and Ferritor 1980). In rural areas especially, where health care is often substandard and specialists unavailable, infant mortality rates are high (Tamblyn 1973). The African American population, in particular, has very high rates of infant mortality that sometimes rival levels in the so called "Third World" (Fuchs 1992). Lack of prenatal care among poor Americans is widely recognized as a major contributory cause of high infant mortality rates.

Many poor counties have no hospital and it is not unusual for county hospitals to close due to financial problems. Patients needing hospital care must make a long and, perhaps, arduous journey. The poor are likely to lack the means of transportation to make such a journey, and the financial means to pay for such things as emergency room visits. It seems likely that in these circumstances routine care is not sought, and that only when medical problems reach a crisis that may be beyond the point of effective intervention do the poor receive health care.

Education

Income poverty is often associated with short duration and poor quality education. A greater proportion of the nonpoor are high school graduates than are the poor (U.S. Department of Commerce 1988). Poor counties often have a small school system that lacks funding because of an inadequate local tax base. Expenditures per pupil are frequently low and the course offerings limited (Zhou and Shaw 1993). These problems often result in a poor quality education. Individual and community attitudes may help to create and compound local problems within education. Education may not be valued by the local community or within the family (Tamblyn 1973). Those who seek education may be scorned. The result is that students leave school at the earliest opportunity and may be barely literate.

Again, a circularity exists between low income and inadequate education. A low wage population may not be able to provide needed funds for a quality education system. Children from low income families who live in a region where low-paying and menial jobs are the norm, are unlikely to value education. A poorly educated populace is unlikely to attract high-paying jobs that might be able to address income poverty, provide a task base for education, and provide motivation to invest in human capital.

Again, it is in rural areas and among the African American population that poverty within education is most profound. Rural children, in general, get less education than their urban counterparts, and the quality of education that is received is lower. The education received by much of the African American population across the rural South is especially inadequate, and can only serve to entrench the income poverty experienced (Task Force on Economic Growth and Opportunity 1969). In addition, the economic possibilities that can be commanded for a certain level of education vary by race. For each given level of education, African Americans earn less than their white counterparts (U.S. Department of Commerce 1988).

Poor education represents an underdeveloped human resource (Bishop 1969). Those who do manage to develop their human capital despite the limitations of the education available, find few avenues through which to apply it. They are exceptions to the norm, and are forced to migrate to areas that offer more opportunity (Smith 1990).

Lack of post-high school education has also recently been associated with poverty. Individuals with low income attend college less frequently than the more affluent, even when they graduate from high school. From 1985 to 1988 only 45 percent of those 18 to 19 year olds who graduated high school who were in the lowest income fifth, attended college. The figure for those in the highest income fifth was 79 percent. Moreover, American colleges are stratified by income. Poor Americans that do attend college are more likely to attend two year public schools than rich Americans. Finally, for those who do enroll in college, attrition rates are much higher for low income students. College graduation is well accepted to be linked to later labor market outcomes, and it seems that poor 18 year olds have difficulty using the avenue of higher education to escape from poverty (Manski 1992/1993).

Housing

Little attention appears to have been paid within the literature to housing quality and living conditions. Housing quality may be substandard in terms of physical structure and facilities. While statistical data may capture inadequacies such as lack of plumbing facilities, overall housing quality is more difficult to assess. Anyone who has driven through a poor section of an American city or through a poor rural area can readily observe poor housing quality. The squalor, the crumbling and dilapidated buildings, the broken windows, and the leaking roofs are readily visually apparent, but capturing such conditions with numerical data is more difficult. Such poor housing must be associated with income poverty, as the poor are forced to rent substandard housing. Even the poor who own their own homes often cannot afford to make needed repairs and must shoulder the financial burdens associated with ownership.

Housing that is in good repair and which possesses such things as complete plumbing facilities cannot always be considered adequate. Overcrowded living conditions may be a result of income poverty. Large families may not be able to afford adequately sized accommodation, and may be living in a small apartment or house. The poor are also sometimes forced financially to live with relatives.

Unlike health and education, the housing dimension of poverty is primarily an outcome of income poverty. However, the nonpoor are able to consolidate their financial advantage through the avenue of home ownership. Mortgages provide substantial tax breaks, and fluctuating property prices in the past have enabled owners at the upper end of the housing market to make large profits.

CHARACTERISTICS OF THE POOR

Who are the poor, and how does the poverty population vary from the general population? It is clear that some groups are more likely to live in poverty than others (Solow 1990). As Axinn and Stern (1988) remark, poverty is etched into age, gender, family, and racial structures. The poor as a group are also distinct from the general population in terms of employment status and characteristics. The characteristics of the poor have not remained static, but have responded to public policy as well as economic and social changes.

Age and Poverty

A major change has occurred in the extent of poverty experienced by the elderly population in the United States. In terms of the population aged over 65 it appears that Johnson's war on poverty was dramatically effective. In 1965, some 33 percent of the elderly were poor, but by 1983, this percentage had fallen to fourteen. This decrease in elderly poor was due partially to Social Security payments and Medicare (Haveman 1987). America's elderly population receive a large proportion of social spending. It should be noted, however, that the transfer payments that have removed many of the elderly from the ranks of the officially poor have done so only marginally. Many of our nation's older citizens have incomes just above the poverty line and so may not be poor by official definition but lead a far from comfortable life in old age (Ruggles 1992).

While there has been a decrease of elderly poor, a very different picture emerges at the other end of the age continuum. Increasingly, the poor consist of children rather than the elderly (Goldstein 1986; O'Hare 1988). In recent years there has been a steep increase in poverty rates amongst children. Poverty rates among all children rose from 16 percent in 1977 to 20.5 percent in 1985. For very young children, and among certain racial and ethnic groups, rates are even higher. In 1985, one in four children below the age of six lived below the official poverty line (Greenstein 1986). Two fifths of very young Hispanic children were poor, as were half young African American children. Poverty among the elderly has decreased, but income has gone down, and benefits have been cut, for families with children (Greenstein 1986; Tobin 1990). Not only is poverty common among children but it is especially persistent (Tobin 1990).

It is clear from census information that age is a factor in the likelihood of experiencing poverty. In 1987 half the poor were either elderly or under the age of eighteen (U.S. Department Of Commerce 1988). Children are constituting an increasing proportion of the poor (Greenstein 1986).

Gender, Marital Status, and Poverty

Women as a group are more likely to live in poverty than are men, and there are more poor women than poor men (O'Hare 1985). Frequently, families where the head of household is female live in poverty, and the poverty population is increasingly composed of such families (Hoffman 1991; Renwick and Bergmann 1993; Rodgers and Rodgers 1993). In 1984, half of all poor families were headed by women, yet female headed households constituted only 15 percent of the total number of families (Jones and Kodras 1990). By 1983, over 40 percent of persons within female headed households were poor, and 35.4 percent of all the poor lived within these households (O'Hare 1985). As

we enter the decade of the 1990s fully half of female headed households
have incomes below the official poverty line (Hoffman 1991). Persons
living within female headed households are predominantly children.
Female headed households often survive on the receipt of welfare
payments such as AFDC (Hoffman 1991). The number of female
headed families has been growing rapidly in the United States, as has
the percentage of children that live in such families (Jones and Kodras
1990).

The immediate reasons for the existence of female headed
families are two-fold. The first reason is divorce. This appears to be an
important factor among white women. The second reason is unmarried
parenthood. Many African American women are unmarried mothers
(Jones and Kodras 1990). Jones and Kodras (1990) offer supporting
evidence for this racial difference. The highest white illegitimacy rate is
found in California, and is 18 percent. In contrast, the highest black
rate, found in Pennsylvania, is 67 percent. Black illegitimacy rates are
on the rise. In 1970, 37.5 percent of African American children were
born out of wedlock; by 1980 this rate was 55.3 percent (Jones and
Kodras 1990). While some may attribute high levels of illegitimate
births among African Americans to some kind of moral decline, many
link the phenomenon to economic causes. Researchers such as Wilson
(Wilson 1978;1987; Wacquant and Wilson 1989), for example, argue
that young black males are so economically disadvantaged that they
cannot, and do not, shoulder the financial responsibilities of marriage
and family. There is a shortage of marriageable African American men
who have a decent job and so can support the luxury of a traditional
nuclear family. It is not only women who are single parents who are
likely to be poor; on average women's income is only 60 percent of
men's (Jones 1990).

Race and Poverty

White Americans constitute a majority of the poor in the
United States (O'Hare 1985), but this disguises the fact that the white
poverty rate is much lower than for other racial groups. Poverty
remains a chronic problem within the African American sector of the
population (Rodgers and Rodgers 1993). Despite poverty rates rising
fastest for young white males from 1973 to 1983 (Axinn and Stern
1988), poverty rates for black Americans remained three times those for
whites (O'Hare 1985). There have been improvements for the African
American population, but the difference between white and black

poverty rates remains large (Axinn and Stern 1988). Non-white families are more likely to be poor and stay poor (Vitelli 1968).

Many African Americans that are poor are members of female-headed families. They are women and their children. However, even for working African Americans the financial prospects are less optimistic than for their white counterparts. The median income for black men in 1983 was only 58 percent that for whites (O'Hare 1985). The mid 1980s still saw over thirty percent of all black Americans living in poverty. Moreover, the percentage of black children in poverty was even more alarming. In 1986 over 42 percent of all black children in the United States were living below the official poverty line. In 1991 it seems that median income for African Americans remained well below that for whites. While white median income was $31,569, black median income was $18,807 (Pollard 1992).

The rural African American population seems particularly prone to live below the official poverty line. Over 40 percent of rural African Americans live in poverty (McCormick 1988). Poverty among African Americans is not limited to low income. As has been previously discussed, African Americans are more likely than other groups to be impoverished educationally (Task Force on Economic Growth and Opportunity 1969; U.S. Department of Commerce 1988). Also, poor health among the African American population is a disturbing manifestation of poverty. For example, infant mortality rates for African Americans are double those for white Americans (Fuchs 1992).

Race remains a major demographic variable associated with varying economic and social deprivation (Dhillon and Howie 1986). There is continued and intense interest within academia in linking the African American experience with economic outcomes. This interest is evident in the extensive literature concerning the urban underclass (Kasarda 1990; Knox 1990), as well as in discussions concerning rural poverty (Dhillon and Howie 1986).

Employment and Poverty

Unemployment is often used as an indicator of poverty and yet unemployment statistics provide a very imperfect measure. Certainly those who are unemployed have a high likelihood of being poor but poverty extends well beyond the officially unemployed population. Unemployment figures fail to include discouraged persons who have ceased to look for work. Moreover, many of the poor are not employable, are not looking for work, and so are not included among those categorized as unemployed. This group includes mothers with small children, children, the sick and disabled, and the elderly (Velasquez 1987).

Just as unemployment is not a good correlate of poverty, nor is employment closely associated with nonpoor status. A growing number of adults in recent years were poor despite working (O'Hare 1985; Goldstein 1986; Anon 1989; Tootle 1989). The 1980s brought an end to any hope that employment would be a security against poverty. Employment was no longer any guarantee of material comfort. In 1986, twelve percent of young workers were poor (O'Hare 1989), and in 1988, 6.7 million full-time workers remained at or near the poverty level (League of Women Voters 1988). In 1985, two million people and their families worked full time all year, but were still below the official poverty line, and working Americans constituted the highest percentage of the poor since records have been kept (Levitan and Shapiro 1987). Working may even mean increased poverty for some, as a job may pay less than previous benefits, and necessitate such expenses as child care (League of Women Voters 1988;Renwick and Bergmann 1993). Today approximately one third of all the poor are part of the workforce.

O'Hare points out that the basic reason for the existence of the working poor is low wages. A full time job which pays minimum wage puts a family of three below the official poverty line. Research indicates that there has been a proliferation of low wage jobs associated with a structural shift within the American economy. There has been movement from a manufacturing to a service economy and the attendant structure of employment has been changing. Most new jobs are within low paying service sectors, and thus may not offer workers sufficient income and benefits to raise then out of poverty (Noyelle and Stanback 1984; Noyelle 1987; Nord and Sheets 1992). In the America of the 1990s, full-time work is no guarantee of an escape from poverty (Jones 1990).

SPATIAL VARIATION IN POVERTY

Poverty statistics tend to focus on nationally aggregated data and, often, social research lacks a spatial perspective (Smith 1972). Poverty research is no exception. Yet, there is an important spatial dimension and organization to United States poverty. The poor in this nation are not dispersed homogeneously in terms of population or place (Peet 1972; Gray and Peterson 1974; Haveman 1987). Poverty appears to be becoming increasingly spatially concentrated (Smith 1987; Eggers and Massey 1991). Poverty is a function of the internal dynamics and the external forces affecting place, and thus spatiality must be a concern in analysis (Chinitz 1969).

Poverty varies spatially depending on the nature of place, and poverty research tends to divide along urban-rural lines. In addition, poverty in the United States historically has varied regionally, and continues to do so today. These two basic spatial elements to poverty are discussed later.

Urban versus Rural Poverty

Poverty is related to size of place (Ford 1969). Perhaps the most fundamental division in the nature of place that affects poverty, is the urban-rural distinction (Hirschl and Rank 1991). Recent focus has been on urban poverty, a more visible and concentrated phenomenon than rural deprivation (McCormick 1988; Smith 1990).

Urban Poverty. The images of poverty that most often confront us from the media are those associated with the city. The urban poor have been the focus of much research (McCormick 1988; Smith 1990; Dudenhefer 1993). Many studies have concentrated on the special groups that tend to make up the urban poor, in an effort to illuminate causality (Dobelstein 1987). If one looks at rates of overall metropolitan poverty, one finds that they are lower than rural rates (McCormick 1988; Hirschl and Rank 1991). However, because of the urban nature of American society, these lower poverty rates represent large absolute numbers of the poor. Similarly, bigger cities often have lower poverty rates than do smaller ones, but yet are home to larger numbers of impoverished Americans (Morrill and Wohlenberg 1971).

Poverty within American cities is not evenly dispersed, but rather is becoming increasingly concentrated both spatially and demographically (Smith 1987; Eggers and Massey 1991). The image of urban poverty is often one of deprivation concentrated in and, largely confined to, the inner city. It should be noted, however, that poverty exists in the suburbs and affects substantial numbers of people.

The spatial and demographic concentration of poverty has led to identification of those affected as an "urban underclass" as noted earlier. There is much debate as to whether "underclass" is an appropriate term, whether such a group exist, and if so what the size of such a group is, and what factors lie at the root of their existence. Despite these debates there is little disagreement about the serious and disturbing nature and trends of urban poverty.

The urban poor are becoming more intensely impoverished (Eggers and Massey 1991; 1992) and geographically concentrated (Smith 1987; Greene 1991; Eggers and Massey 1992). Poor urban Americans are increasingly found in segregated deprived neighborhoods within the central city (Kasarda 1990). Not only are the urban poor becoming more geographically concentrated but they are also becoming increasingly socially and occupationally isolated. Such isolation results in poverty which is much more persistent and intractable (Eggers and Massey 1992). Another characteristic concerning urban poverty, about which there is little disagreement, is that it is concentrated in the non-white population (Kasarda 1990).

Rural Poverty. Rural poverty has faded from the limelight and the rural poor have become largely invisible, with the recent emphasis on the urban poor (McCormick 1988; Smith 1990; Dudenhefer 1993). However, rural poverty appears chronic and remains substantial (Mertz 1978). Small communities and nonmetropolitan areas often have inadequate institutional systems and lack economic resources (Ford 1969), and many of the poor live in small towns (McCormick 1988). Poverty rates are substantially higher in nonmetropolitan as compared to metropolitan areas, and the gap is widening (Brunn and Wheeler 1971; Tootle 1989; Hoppe 1991). In 1970, 30 percent of the U.S. population was considered rural, yet rural areas were home to half the nation's poor. Non-metropolitan poverty rates at that time were double those for metropolitan areas (Tamblyn 1973). Conditions and prospects for America's rural areas do not seem to have improved. Rural industry is in decline and a majority of new job opportunities are in metropolitan areas. The rural United States is caught in a chronic recession (McCormick 1988).

Poverty rates in some rural counties are extremely high, and it seems poverty is the norm. Tunica County, Mississippi provides an example. In Tunica County over 50 percent of the population lives below the official poverty line. Two thirds of the housing in the county is substandard, and half the population survives on transfer payments. The county provides an extreme example of a limited economic base which is typical of rural areas. Only one manufacturing plant is located in the county, and this Pillow-Tex/Wal-Mart plant can hardly be considered high-tech (Smith 1990). Rural poverty is long-standing, despite the affluence of the United States. It seems that rural poverty is particularly intransigent and is rooted in historical, economic, social, and cultural dynamics of the local community (Reul 1974; Fitchen 1981).

Rural areas and rural residents suffer from a host of problems. Despite media focus on the social problems of the inner cities, such problems also occur in rural areas. Alcohol abuse and illegal drug use are common, as are family problems (McCormick 1988; Smith 1990). Rural residents are more likely than their urban counterparts to suffer from chronic disease and disability. This is partly due to environmental and work conditions, but also to out-migration of the young and healthy. Rural areas have limited and diminishing economic opportunity, and many jobs are low paying. To compound this problem some costs faced by rural residents in their daily lives may be higher than in urban areas. Food, especially in isolated rural areas, may be more expensive and the costs of transportation higher (Smith 1990). One has only to observe grocery and gas prices in more remote rural areas to appreciate that some elements of the cost of living are higher than in much of the rest of the nation.

Nonmetropolitan areas receive a disproportionally low share of federal outlays even when population size is adjusted for. Less money is channeled into such areas as housing, job training, education, and health provision. Lack of health care, and low health care standards are a major social deficit in rural areas (Tamblyn 1973). High infant mortality is a disturbing result of this inadequate health care. In America's poorest 320 rural counties, infant mortality rates are 45 percent higher than the national average. Often health care is a luxury, and by necessity is only sought during periods of crisis (McCormick 1988). Education in rural areas is also often substandard (Task Force on Economic Growth and Opportunity 1969). Schools are frequently small, and expenditures per pupil are low (Zhou and Shaw 1993). Funds for adult education are often meager. In addition to problems of poor health and inadequate education, rural residents often face poor living conditions. Substandard housing is most frequently to be found in rural areas (Tamblyn 1973; Ropers 1991).

The rural poor use the welfare system less than do their urban counterparts. Low participation may be the result of several factors. One reason for low participation may be difficulty of physical access to services. Another possible factor is lack of information. In addition, cultural and psychological factors cannot be discounted. The use of such things as food stamps may bring social disapproval (Hirschl and Rank 1991). This cultural pressure may be a powerful force (Reul 1974). Also, personal pride may be a factor in low participation in welfare programs. To accept food stamps or transfer payments may be perceived by individuals to be an admission of personal failure (McCormick 1988). In many rural areas people have a long tradition of being fiercely independent, and such independence brings respect and approval from others. Since rural residents tend to be part of a part of a well-developed network of family and friends social approval is important. The process of applying for welfare is not a pleasant one, and may often require levels of literacy that potential recipients lack. The welfare available may be perceived as not worth the humiliations and invasions of privacy that are a part of initiating and maintaining benefits.

Regional Dimensions of Poverty

Beyond the urban-rural dimension in the spatial characteristics of poverty, poverty rates show broad regional variation in the United States. Regional disparity is endemic in the United States (Smith 1979). In 1972, Smith identified regions at varying levels of well-being. In his classification the relatively affluent northern United States contrasted sharply with a region of social deprivation that spanned the southern states from Texas to Virginia (Smith 1972). The most economically and socially deprived region in the United States is the Old South (Smith 1972; Mertz 1978; Agnew 1987; Smith 1987).

While poverty did decline fastest in the South during the liberal era of federal policy, the region was still home to 40 percent of the poor in 1972 (Plotnick and Skidmore 1975). Porter (1989) identifies 206 counties that remained in the bottom fifth nationally, based on per capita income, over the years 1950, 1959, 1969, 1979, and 1984. All but eighteen of these counties were located in the South (Porter 1989). In the rural South, poverty amongst children, especially black children, is epidemic. Sixty-eight percent of young southern black families were poor in 1986, as were 57 percent of southern black children (O'Hare 1988). Over half the nonmetropolitan poor live in the South, and over 40 percent of these impoverished Americans are black.

In general, the southern link with poverty appears to be the prevalent rural lifestyle, with its limited tax base and range of economic activity (Peet 1972). There has been a decline in agriculture and resource linked industries and in routine manufacturing on which the South has depended. In addition, the region has lost many of its most valuable residents with the out-migration of the best educated and most skilled (O'Hare 1988). The history of the region cannot be discounted as a factor contributing to present conditions. Reul (1974) points to exploitation by the North, the legacy of slavery, the Civil War, Reconstruction, as well as paternalism and dependency as important historical dynamics.

Conditions in the South have improved in recent decades. As recently as 1960 half the population of the region had only an elementary school education (Reul 1974). Within the South as a region today, there is great variation in poverty levels (Smith 1987). The persistence of poverty within some southern counties points to the existence of cores of poverty within the rural South despite overall economic improvements (Mertz 1978; Shaw 1992, 1990b). The research of both Tamblyn (1973) and Brunn and Wheeler (1971) identified Appalachia, the Ozark Plateau, and the southern black belt to be deeply impoverished areas within the South. By 1990 it seems that little had changed in the geography of southern poverty. Jones and Kodras (1990) discussed the location of areas that suffer from extreme poverty. One poverty core is located on the Texas gulf coast and another in the Appalachians of eastern Kentucky and southwest West Virginia. A third area of intense poverty mirrors the distribution of the African American population and cuts a swath across the lower Mississippi Valley (Yazoo Basin), central Alabama, and the southeastern piedmont and coastal plain. Other areas where poverty is endemic exist where the Native American populations dominate (Jones and Kodras 1990).

Much of the work concerning the rural dimensions of poverty in the United States was conducted in the 1970s when interest in poverty research was at its height. Since that time, regional discussions of poverty have been generalized. It seems that little systematic research has been conducted on regional variation in poverty levels, or on regional variation in the nature of poverty. This lack of spatial analysis of poverty appears to represent an important area that has not been fully investigated within geographic research (Giarratani and Rogers 1991).

V

Analysis of Poverty's Distribution and Characteristics

INTRODUCTION

The analysis of poverty's distribution and characteristics in the United States presented in this chapter focuses primarily on the year 1980. This focus is adopted primarily because 1980 can provide a low level "baseline" with which to contrast 1990 poverty distribution and characteristics in subsequent temporal analysis. The distribution, extent, and nature of poverty in 1980 represents conditions of deprivation after some fifteen years of spending during the late 1960s and 1970s in an effort to alleviate poverty.

THE SPATIAL DISTRIBUTION OF POVERTY AND IDENTIFICATION OF POVERTY CORES

Data

In order to investigate the spatial distribution of the poor within the United States, and identify poverty regions, some overall measure of poverty must be selected. For most governmental and reporting purposes, the measure of poverty used is the head count of persons or families whose income falls below the official poverty line (Orshansky 1969; Oster, Lake, and Oksman 1978; Rodgers 1991). As has been discussed, there exist dimensions to poverty that lie beyond mere lack of income. However, although many nonincome measures of poverty have been recognized (Vitelli 1968; Ferman, Kornbluh and Haber 1969; Cowell 1977; Goldstein and Sachs 1983; Harrington 1984; Levitan and Shapero 1987), and several poverty characteristics identified, lack of income is still seen as fundamental to poverty in the

United States (Peet 1972; Harvey 1985). In a market economy, such as exists in America, it is money income that primarily commands access to health care, quality education, and decent housing. It is income that enables or constrains individuals and families in their participation in what might be considered "normal" activities within American society. Thus, some measure of income is almost invariably used to assess the poverty being experienced.

Recent research has focused on the refinement of income measures of poverty. Several different, sometimes complex, poverty indices have emerged from the econometric literature which seek to improve upon a simple count of persons who fall below some designated income level (Rodgers 1991). However, these sophisticated measures of income, share a common problem with the more frequently used per capita income measure: no adjustment is made for spatial variations in cost of living (Ford 1969; Goldstein and Sachs 1983), which may vary considerably at both a regional and county level (Cowell 1977; MacDonald 1969). This study admits the practicality of using income as a measure of poverty and also accepts lack of income to be fundamental to poverty in the United States. Since lack of income is widely accepted as being the essence of poverty, an income variable will be used in this study to investigate the spatial distribution of poverty. The lack of spatial sensitivity of commonly used income measures is, however, problematic in this as in any geographic study.

Housing costs account for a large share of living expenses, and constitute the costs that vary the most spatially (Bureau of National Affairs 1988). The substantial spatial variation in housing costs is clear if one views the median housing costs reported by the Bureau of the Census at the county level. The costs reported include mortgage payments, real estate taxes, property insurance, cost of utilities, and fuel costs. In 1980 median monthly housing costs in Pitkin County, Colorado were $739 compared to $180 in Kemper County, Mississippi; in both counties the official level of income poverty was identical (U.S. Department of Commerce 1983).

The lack of sensitivity to variations in the cost of living, inherent in traditional measures of income, makes a more realistic measure of income which adjusts for basic living costs desirable. In an attempt to address this problem of spatial insensitivity, an income variable is developed that adjusts for variation in living costs, and thus, more accurately captures the probable deprivation being experienced.

As was previously stated, it is housing costs that show large spatial variation; other costs, such as food, are much less variable geographically and are hence much less influential in determining the variation in buying power of income. The income variable used will, thus, be adjusted for varying housing cost.

The income variable (INDEX) to be used in this study, for both 1980 and 1990, will be calculated as follows:

$$\frac{\text{Median household income - median housing costs}}{\text{Average number of persons per household}}$$

Housing costs, which include mortgage payments, real estate taxes, property insurance, utilities and fuel costs, are incurred at a household level and are relatively independent of household size, hence it seems appropriate to deduct such costs from household income. The income left to a family after housing costs have been deducted is representative of money that would be available were that family sharing in the general level of material well-being in their county of residence. In this way living cost variation is, in large measure, allowed for. Since other costs, besides housing, are usually incurred on an individual level the money income remaining after deduction of housing costs was divided by the average household size. It is this per capita income that remains after deduction of housing costs that will be used in this study. As is the case with other per capita income data, these statistics are an average for the areal units employed - in this case counties. These averages give no indication of the degree of inequity within a county, but give only an indication of the economic condition of the average resident. Nor do such data give any indication of the absolute numbers of people involved.

The county level data on income, housing costs, and household size used to develop this income index were obtained via mainframe computer access of the magnetic tape version of the Bureau of the Census County and City Data Book (U.S. Department of Commerce 1983), and also Summary Tape File 3 from the 1990 U.S. Census. Data were obtained for all 3109 county or county equivalents contained within the forty eight conterminous U.S. states in 1980, and for the 3111 county or county equivalents that existed in 1990.

Methodology

In order to assess the distribution of poverty and identify poverty regions within the United States the adjusted Per Capita Income (INDEX) was mapped at the county level for both 1980 and 1990. This was accomplished by use of the computer mapping/geographic information system package ATLAS*GIS. This package has the ability to match provided data to an areal unit and to produce choropleth maps. In order to create such maps, the data must be matched to the appropriate unit or polygon. This matching is accomplished by the use of FIPS (Federal Information Processing Standard) codes which are assigned to each county within the numeric data file and to each county polygon within the ATLAS*GIS boundary file. The county boundary file available within ATLAS*GIS was not contemporary with the 1980 census data used in this study and so two county polygons, one in Arizona and one in New Mexico, have missing data. No county polygons have missing data when the 1990 census data is mapped.

Subsequent to matching the data to the appropriate geographic unit, the INDEX was mapped for 1980 using five classes defined by the standard deviation of the data. For comparative purposes a similar map of unadjusted Per Capita Income was produced.

A map was also generated using the poverty density statistic for 1980 - that is, the number of persons below the official poverty line per square mile. The previously produced maps reflect only the poverty rate being experienced within a county, and give no information concerning the absolute numbers of people involved. The poverty density map is provided to give a sense of where the poor are located within the United States in terms of absolute numbers rather than average condition. While the poverty rate may be quite low in one county and high in another, the number of people experiencing poverty within a given areal extent may often be quite different than such differential poverty rates suggest. Since the poverty density statistic does not approximate a normal distribution, mapping by standard deviation results in an undifferentiated map. Consequently, this variable is classed into 7 quantiles for the purposes of mapping. In total, seven maps were produced to aid in the investigation of the spatial distribution of U.S. poverty, and in the identification of poverty regions.

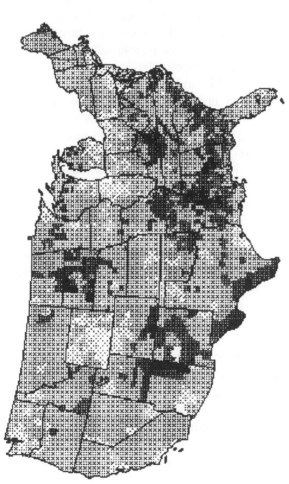

Figure 3. Spatial Variation in the Income Index in the U.S. - 1980

Results and Discussion

Figure 3 presents the mapped income index and reveals that a distinct spatial pattern characterizes United States poverty in 1980. In contrast, while mapping the unadjusted Per Capita Income at the county level does illustrate broad regional variation in poverty levels, it fails toidentify distinct cores of poverty. The relative poverty position of some counties is substantially influenced by the housing cost adjustment to Per Capita Income. This is illustrated in Appendix A in which the poorest 5 percent of counties in 1980, as defined by adjusted Per Capita Income, are also ranked by unadjusted Per Capita Income. It is clear that the poverty ranking of some counties is largely unaffected by the adjustment; the adjusted and unadjusted rankings for Starr County, Texas are 3 and 6 respectively. For Cumberland County, Kentucky the rankings are 111 and 109, and for Clay County, West Virginia the adjusted and unadjusted rankings are 141 and 143. However, most counties exhibit marked changes in ranking when Per Capita Income is adjusted for variation in housing costs. Some, such as Costilla County Colorado (ranked 6 and 1712), Bronx County, New York (ranked 81 and 2051), and Hudspeth County, Texas (ranked 140 and 2720) are identified as being relatively much poorer when the adjusted rather than unadjusted Per Capita Income is used to assess poverty. Other counties such as Van Buren County, Tennessee (ranked 774 and 7), Catron County, New Mexico (ranked 405 and 15), and Rich County, Utah (ranked 1562 and 16), which are classed among the poorest of counties by unadjusted Per Capita Income are shown to be relatively affluent when housing costs are considered.

Using adjusted per capita income in 1990 as opposed to per capita income in 1990 influences the relative poverty position of counties in a similar way as that described for 1980. While some counties are similarly ranked in terms of poverty by both measures, others show substantially altered poverty ranking. Poverty as measured by Per Capita Income is a relatively widely dispersed phenomenon in 1980, especially through the southeastern United States. This broad poverty region extends east-west from the Carolinas to Texas and north-south from West Virginia to northern Florida (Figure 4). The spatial pattern of poverty as expressed by per capita income is consistent with patterns identified by other researchers. Smith (1972) identified a swath of poverty that extended across the southern United States from Virginia to Texas. The impoverished South was contrasted with a much more affluent North (Smith 1972). Mapping per capita income for 1980

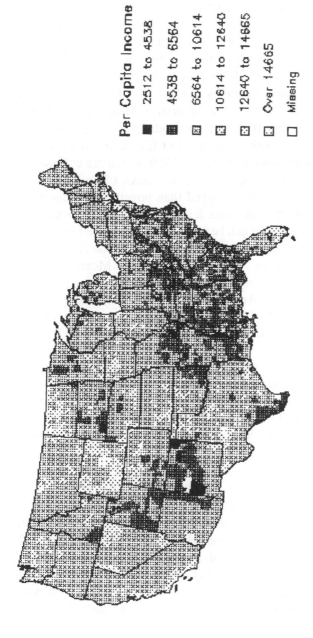

Figure 4. Spatial Variation in the Unadjusted Per Capita Income in the U.S. - 1980

indicates that the North-South contrast identified by Smith has persisted into the decade of the 1980s.

Other researchers who went beyond identification of broad spatial patterns of poverty generally agreed upon the existence of some spatial cores of poverty. Tamblyn (1973) and Brunn and Wheeler (1971) identified Appalachia, the Ozark Plateau, and the southern black belt to be cores of poverty. Jones (1990) identified the southern black belt, the Appalachians of eastern Kentucky, the Texas gulf coast, and regions of the southwest with concentrations of Native American populations to be areas where poverty is a common condition. The spatial pattern of poverty defined by housing adjusted Per Capita Income reveals five major core areas of poverty existed in 1980, which are consistent with portions of this previous research. These cores of poverty are much more spatially distinct when mapped than poverty areas discussed by researchers previously mentioned. It is in these quite distinct cores, that poverty at the county level can be said to be concentrated in 1980.

The first poverty core is located in the region encompassed by southern Utah, northeast Arizona, western New Mexico, and south-central Colorado. This poverty region is located in an area with high percentages of Native American population (U.S. Department of Commerce 1983). Indeed, much of this poverty core, including the intensely impoverished Apache County, Arizona, is located on reservation land.

A second area of intense poverty exists throughout central South Dakota extending into south-central North Dakota. This poverty core is also in an area where Native American population concentrates (U.S. Department of Commerce 1983). Again, much of this poverty core, such as Corson, Shannon, Todd, and Ziebach Counties is reservation land.

A third distinct poverty area extends the entire length of the border of Texas with Mexico. This appears to be in keeping with the research of Jones and Kodras who link poverty status to Hispanic ethnicity (Jones and Kodras 1990). Of the 13 Texan counties with the highest percentage of population who are of Hispanic origin, 11 are contained within the identified poverty core. The most intensely impoverished counties of Maverick, Presido, Starr, and Zavala have approximately 92 percent Hispanic population compared with 21 percent for Texas as a whole (U.S. Department of Commerce 1983).

A fourth broad area of poverty can be seen to lie along the Mississippi River from Louisiana to southern Illinois; this poverty region also extends across much of Mississippi as well as across the southern coastal plain from Alabama to North Carolina. The region identified supports the comments of other researchers (Smith 1972,1979; Brunn and Wheeler 1971; Jones and Kodras 1990) who suggest that the concentration of poverty across the South mirrors the distribution of the rural African American population. The most intensely impoverished of the counties contained within this core lie along the Mississippi River from central Arkansas and northern Mississippi to northern Louisiana and southern Mississippi, and in southern Alabama. All these counties, with the exception of West Carroll County, Louisiana, have substantially higher percentages of African American population than the averages for their state. In these most severely impoverished counties African Americans are a majority of the population (U.S. Department of Commerce 1983).

A fifth core of poverty which is particularly geographically coherent spans most of eastern Kentucky and extends into northern Tennessee. In contrast to the four previously identified cores of poverty this area has no concentration of minority population. Instead, the population of this region is overwhelmingly white and of Anglo-Saxon origin and thus is not ethnically or racially distinct from the American majority.

In addition to the five major cores of poverty discussed, two less intense but distinct areas of poverty seem identifiable from Figure 3. The first extends across the Ozark Plateau of southern Missouri and northern Arkansas, in an area which is similar in many respects to Appalachia. The second is located in eastern Oklahoma. The impoverished areas of Oklahoma, while they are not reservation land, do appear to be home to relatively large numbers of Native Americans (U.S. Department of Commerce 1983).

The concentration of poverty in relatively few core areas is further illustrated by Table 1. This table reveals that the poorest 50 counties are located in just 11 states; moreover, these same 11 states account for 124 of the poorest 150 counties. Two states alone (Mississippi and Kentucky) encompass 20 of the poorest 50, 34 of the poorest 100, and 50 of the poorest 150 counties. It is the poverty areas centered in Mississippi and Kentucky that can be said to represent the cores of the most intense poverty within the United States (Table 1). Figure 5 displays the poorest 5 percent of counties in 1980. This map of the poorest 5 percent of counties identified in Appendix A vividly shows that extreme poverty in America is a distinctly spatial

Table 1. Representation of States in the Poorest 150 Counties in 1980

State	Number in 50 Poorest Counties	Number in 51 - 100 Poorest Counties	Number in 101-150 Poorest Counties
Kentucky	11	8	6
Mississippi	9	6	10
S. Dakota	6	6	3
Alabama	6	2	3
Texas	5	4	
Arkansas	4	4	2
Louisiana	3	2	3
Georgia	2	7	4
Colorado	2		
Tennessee	1	2	2
Arizona	1		
New Mexico		4	
Missouri		2	3
N. Dakota		1	2
Nebraska		1	
New York		1	
Oklahoma			3
Florida			2
N. Carolina			2
W. Virginia			2
Michigan			1
S. Carolina			1
Utah			1

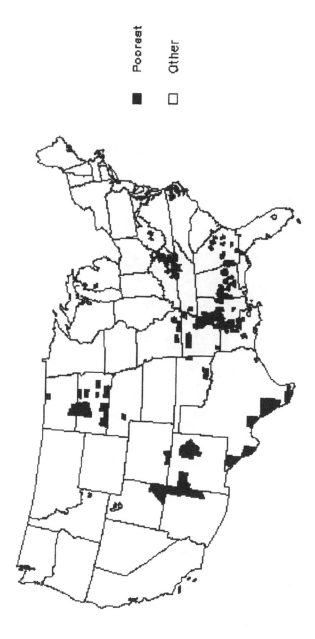

Poorest

Other

Figure 5. Spatial Distribution of the Poorest 5 Percent of Counties - 1980

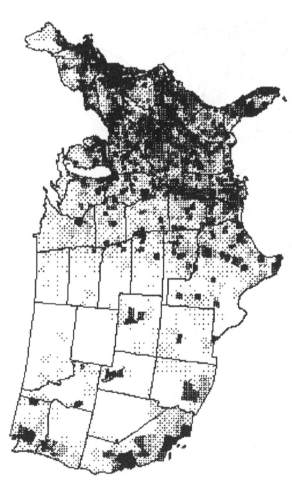

Poor per sq mile

☐   0 to 1.0

▣   1.0 to 3.0

▣   3.0 to 4.7

▣   4.7 to 7.1

▣   7.1 to 10.4

▣   10.4 to 20.0

■   Over 20

Figure 6. Poverty Density in the U.S. - 1980

phenomenon. The poorest counties are far from evenly dispersed and congregate in very distinct cores.

Figure 6, which displays poverty density, reveals a spatial distribution of the absolute numbers of the poor that might be expected. In general the poverty density within the United States reflects the general distribution of the population. The eastern half of the nation and the west coast region have much higher poverty densities than do the Rocky Mountain and Great Plain states. Large cities such as Seattle, San Francisco, Los Angeles, Salt Lake City, Denver, Dallas, Houston, St Louis, Chicago, Minneapolis, Pittsburg, Atlanta, and Miami can all be clearly identified as the location of large numbers of poor Americans. However, looking beyond the mirroring of the general distribution of the population some interesting observations can be made.

A ribbon of relatively high poverty density can be seen to extend along the Mississippi River through northern Mississippi/southern Arkansas. An area of high poverty density can also be seen in southeastern Kentucky. Neither of these areas with a high poverty density are the site of large cities or are otherwise particularly densely populated, yet they are included in the highest poverty density quantile. Thus, it seems that the two areas not only exhibit high rates of poverty, but also are home to large absolute numbers of the poor.

It is clear from the results presented above that this research indicates that poverty in America is distinctly a spatially varying phenomenon. The poor are not evenly distributed across the conterminous United States but instead are predominantly to be found in particular locations. The distribution of the poor cannot be fully understood by generalization on a broad regional basis. Beyond any doubt, poverty is more commonly a southern than a northern condition, as other researchers have pointed out (Smith 1972; 1979; Dudenhefer 1993). However, merely to contrast the poor South to the affluent North is to miss important spatial dimensions of poverty. The South, in common with other areas of the country, is not homogeneously poor. Rather, it is in very specific and clearly bounded areas within the South that the southern rural poor are concentrated. Thus, it seems that any approach that seeks to address the problems of poverty by the thrust of broad regional development plans neglects the spatial evidence of the poverty experience. Some researchers characterize the economic problems of the country, and of its population, in terms of regional dynamics. Often cited is the decline in manufacturing in the Northeast and the rise of other regions such as the South based on service activity (Noyelle and Stanback 1984; Agnew 1987; Noyelle 1987). While these changes within the American economy are clear, they serve mainly to explain broad trends. They appear to do little to explain the location of the poor in distinct cores.

Other researchers have gone beyond characterization of broad regions and have suggested that cores of poverty exist (Brunn and Wheeler 1971; Jones and Kodras 1990). This research strongly confirms this contention. The southwestern poverty core spanned parts of Arizona, Colorado, New Mexico, and Utah and included an extensive geographic area in 1980. Brunn and Wheeler (1971) mapped the poorest one thousand counties by median income for 1959, and this poverty core could be clearly seen as an island of relative poverty in the western United States. It is again remarkable to observe the persistence of the poverty core that was apparent in the Dakotas in 1959 (Brunn and Wheeler 1971). This poverty core remained clearly visible throughout central South Dakota and south-central North Dakota in 1980. A third distinct area of poverty was identified in Texas in the counties that hug the Mexican border. In contrast to the previous two cores, this ribbon of poverty was not apparent in the pioneer work conducted by Brunn and Wheeler (1971). Instead, much of east Texas appeared to be part of a more general area of poverty that spanned much of the South. The border poverty core identified in this research by mapping the income index is clearly delineated. All Texas counties that border Mexico are part of this poverty core. Despite lack of confirmation of the existence of this core by previous research, it appears to have persisted throughout the 1980s. A fourth area of poverty identified is centered along the southern reaches of the Mississippi River but extends across much of the southern coastal plain. This poverty core is less generalized than the extensive area of southern poverty identified by previous research (Smith 1972; Agnew 1987). This poverty core has both a north-south and an east-west axis. The north-south axis extends down the Mississippi River valley from southern Illinois to the Gulf coast. In 1980 the east-west axis spanned the southern coastal plain from Louisiana to North Carolina. These two axes intersect in northwest Mississippi, southeast Arkansas, and northeast Louisiana. At this intersection lies a large node of intense poverty. A fifth core of poverty was identified as located in the central Appalachians in 1980. This area has long been recognized as deeply impoverished. Researchers unanimously agree on the existence of this poverty core and on the intensity of impoverishment experienced there. The central Appalachian poverty core can be seen on the Brunn and Wheeler (1971) map of 1959's poorest counties.

THE CONCOMITANTS OF POVERTY AND CHARACTERISTICS OF THE POVERTY POPULATION

Variables

Selection of the variables used in the investigation of the concomitants of income poverty, and the characteristics of the poverty population was based on previous research. This research suggests that the concomitants of income poverty fall in the broad areas of education, housing, health, occupational structure, and residential status. Thus, appropriate variables were selected to represent these concomitant dimensions of poverty. Table 2[1] presents the definitions of the variables used.

The level of educational attainment of county residents is represented by the variables HSCHOOL and COLLEGE. It is anticipated that as the percentage of the adult population that have completed high school rises, so will income. Thus, HSCHOOL is expected to be negatively associated with income poverty. A similar relationship is anticipated to exist between COLLEGE and poverty. While HSCHOOL and COLLEGE do represent educational level reached, it should be noted that they do not capture the quality of the education received. No measure is available for the entire nation at the county level which gives a view of the actual skills and accomplishments of county residents that would represent the reality of the human capital conferred by both formal and informal education.

The quality of housing is explored using the variables PLUMB and OVERCR. Both the percentage of housing that lacks complete plumbing and the percentage occupied by more than 1.01 persons per room is expected to be negatively related to income and positively related to income poverty. The variables PLUMB and OVERCR give some sense of the physical facilities offered in housing units, as well as the level of overcrowding that may be experienced in what may otherwise be adequate housing.

While a variable such as infant mortality which directly indicates the health of the population would be desirable, no such

Table 2.	The Study Variables Utilized
VARIABLE	**DESCRIPTION**
HSCHOOL	Percentage of persons 25 and over who have completed 12 years of school or passed a high school equivalency examination. 1980 and 1990.
COLLEGE	Percentage of persons 25 and over who have completed at least 16 years of school. 1980 and 1990.
PLUMB	Percent of housing units that lack complete plumbing or share plumbing facilities. 1980 and 1990.
OVERCR	Percent of housing units occupied by more persons than there are rooms. 1980 and 1990.
PHYSIC	Number of physicians per 100,000 resident population as of 1st April 1980.
HOSP	Number of hospital beds per 100,000 resident population as of 1st April 1980
LABMAN	Percent of the civilian labor force employed in manufacturing in 1980.
LABRET	Percent of the civilian labor force employed in wholesale and retail trade in 1980.
LABPROF	Percent of the civilian labor force employed in professional and related services in 1980.
COMMUT	Percent of workers working outside their county of residence. 1980 and 1990.
URBAN	Percentage of persons living in incorporated and unincorporated place of over 2500 population, and those who live in urbanized areas. 1980 and 1990.
WHITE	Classified as white by race or nationality indicated by individuals. 1980 and 1990.
BLACK	Classified as black by race or nationality indicated by individuals. 1980 and 1990.

measure is suitable for use at a county level. Infant mortality is a commonly used indicator of general health and well-being. However, at the county level it appears that the sample size of births and deaths is insufficient to calculate a meaningful infant mortality rate in some counties. In counties with few births, statistically insignificant changes in the number of infant deaths may greatly affect reported infant mortality rates. Access to health care, rather than a direct measure of health, is used in this study. The variables selected are PHYSIC and HOSP. The number of physicians and the number of hospital beds per 100,000 county residents are both expected to be positively related to income and negatively linked to income poverty.

Occupational structure is represented by three variables LABMAN, LABRET, and LABPROF. Income is anticipated to rise with the percentage of employed persons who work in manufacturing, since workers in manufacturing jobs tend to be relatively highly paid. The percentage of workers whose job is within wholesale or retail trade is also expected to be positively associated with high income and negatively associated with poverty. The extent of wholesale and retail is closely associated with urbanization and thus relatively diverse opportunities. Despite the literature which points to a concentration of low-wage jobs within the service sector not all are low paid. Low-wage jobs are counterbalanced by some very highly paid jobs within wholesaling and retailing. The percentage of those employed who work in professional and related services is also expected to be negatively linked to poverty and positively associated to relatively high income. Some potential problems should be noted within these variables that represent occupational structure. While these variables do give a broad picture of occupational structure within a county, the subdivisions used are crude. For example, these variables cannot distinguish between those who may work within large manufacturing plants producing sophisticated products, and those who work in small plants producing a low value-added product. Similarly, the wholesale and retail trade category is extremely broad and contains high internal variability.

Related to the occupational structure by sector and opportunity within a county is the access to work in counties besides those in which people live. The percentage of employed persons who work outside their county of residence (COMMUT) is anticipated to be associated with high income. This aspect of employment structure may be viewed as an expression of access, outside opportunity, and lack of internal opportunity all of which are linked to the availability of jobs.

Those who are mobile enough to take advantage of opportunities outside their county of residence or who are affluent enough to choose to reside away from their place of work are less likely to be poor than other Americans. It should be noted that this variable is a less than ideal measure of labor mobility. Counties in the eastern United States tend to be small while those in the West tend to be larger. As a result the same travel distance may be more likely to involve crossing a county line in the East. In addition, for residents who live close to the boundary of their county crossing the county line may require only a short trip or a journey to the closest urban center. Despite these problems, given the lack of actual journey to work data, COMMUT is the best variable available.

Finally residential status, that is rural versus urban location, may be a concomitant of income poverty. The variable URBAN is the percentage of a county's residents that are classified as urban residents. The term "urban," as used by the U.S. Census Bureau, encompasses places with widely divergent qualities. An urban place is deemed to be any with over 2,500 residents. Thus, both a person who lives in a small town of 2,500 and one who lives in a large metropolis such as New York are said to be urban residents. It is apparent that the lived experience will be vastly different for these two individuals, and that a resident of a small town might be more appropriately classified as rural. Despite these problems, the variable URBAN is useful for drawing a broad distinction in terms of residential status between town and country.

The characteristics of the poverty population are generally established to vary from the general population in terms of race, age, gender, family structure, migrational tendency, and employment status. Table 2[1] presents the definitions of the variables used.

Links between poverty and the racial/ethnic composition of the population are captured by the use of four variables, WHITE, BLACK, INDIAN, and SPANISH. The percentage of the population who report themselves to belong to a minority racial or ethnic group is anticipated to be associated with high levels of poverty and relatively low income. Native Americans, African Americans, and those of Spanish ethnicity may come to the job market with weak skills that are a product of poor education and family circumstance. Once in the job market they may well face discrimination. In contrast, the majority white population are more likely to have benefited from a quality education and to be treated preferentially within the job market. Thus, it is expected that the percentage of white population will be positively associated with income. Since all racial data rely on self-reporting some inaccuracy may occur. A particular problem lies in the existence of an "other" census category. In particular, many individuals that might otherwise be classified to be of Spanish origin place themselves in the "other" racial category. Thus, the methodology and categories employed within the U.S. census may affect the accuracy of data relating to the racial composition of the population.

The age structure of county population is represented by four variables, YOUNG, OLD, TEENS, and DEPEND. A population that includes a high percentage of either young or old individuals is expected to be associated with low adjusted per capita income. Similarly, high dependency rates can be expected to place a heavy burden of non-productive individuals on a county.

Various characteristics important within the poverty population relate to gender and family structure; PERFEM, TEENBIR, and LABFEM represent these characteristics. The percentage of families that are female headed and the percentage of births to mothers under 20 years old are anticipated to be positively associated with poverty and negatively related to income. Women who are head of a household or who are teenage mothers have heavy family responsibilities and may be unable to work. Such families may depend upon welfare and so are unlikely to enjoy high income. In contrast, female participation in the labor force is expected to be linked to relatively high income. A women working is likely to bring additional income to a family where two adults are present. Where the women is sole breadwinner, wages from work are frequently higher than income available from other sources. If the only jobs available result in lower income than can be obtained from such things as welfare, the choice may logically be made not to work. Thus, female participation in the workforce may, to some extent, reflect the quality of jobs available.

Migrational tendency is another characteristic that appears to be linked to poverty. Poor areas often lose population when people seek opportunity elsewhere, while more affluent areas tend to gain population. Thus, population change from 1970 to 1980 (POPCH) is anticipated to be positively related to adjusted per capita income. The greater the percentage growth in population in a county the more likely it is that the county is booming economically. Conversely population loss or low rates of population growth are likely to be associated with economic stagnation and hence poverty.

A final characteristic of the poverty population which is of importance relates to occupational status. Thus, UNEMPL, the unemployment rate, and LABPART labor force participation are used as variables. Unemployment, it is clear, is likely to be associated with poverty. Conversely, participation in the labor force is necessary for most Americans if they are to command a decent income. Some problems exist with the use of unemployment data. The Bureau of the Census defines unemployed persons as those civilians 16 or over who did not work and did not have a job during a designated reference week. Unemployed persons must have been looking for work during the previous four weeks and be available to accept a job. Individuals who had not worked during the reference week and who were waiting to be recalled to a job from which they had been laid off were also included as unemployed (U.S. Department of Commerce, 1983). One problem is that those discouraged from looking for work, often the most chronically unemployed, are not captured in the data. Similarly those who are unemployed for a very short period, coincidentally during the reference week, are counted as unemployed. Another problem may lie in the choice of the reference week used. The reference week used in calculation of this variable was in the latter half of March, a time when spring agricultural activity may be utilizing substantial amounts of temporary labor. It seems likely, then, that agricultural areas may show relatively high levels of employment reflective of seasonal levels of employment, rather than the average situation within a county.

LABPART is the number of individuals counted as part of the civilian labor force expressed as a percentage of the population between 18 and 65 years old. This variable is included to capture those people who are not part of the labor force, yet are of working age. These include people discouraged from looking for work and thus not categorized as unemployed, or those who may substitute informal activity for formal employment. It also includes students and those who by choice or necessity are homemakers.

County level statistics for the variables were once again obtained via mainframe computer access of the magnetic tape version of the Bureau of the Census County and City Data Book (U.S. Department of Commerce 1983). Data were obtained for all 3109 county or county equivalents contained within the forty eight conterminous U.S. states in 1980.

Methodology

The interrelationship between poverty variables and their association with the income index was explored in three steps.

1. A Pearson Product Moment Correlation matrix was generated to make a preliminary exploration of the relationship between the socioeconomic variables selected for use in this study. It is expected that these variables will, in general, be highly correlated. The correlation matrix is also used to indicate the simple relationship between the dependent variable INDEX and the independent variables. Table 3 presents the correlation matrix for the twenty five independent variables in addition to the dependent variable, housing adjusted per capital income.

2. Principal Components Analysis was employed to group the array of independent variables into categories. Within each category or factor the variables can be said to be interrelated and each factor represents a dimension contained within the data. Principal Components Analysis is used in this research primarily for two reasons: (i) to overcome the expected multicollinearity within the data that would otherwise be a problem in subsequent analysis; and (ii) to reduce the number of variables in the data set by identifying latent dimensions within it. The analysis may, thus, be used directly to illustrate the concomitants of income poverty, and to identify the characteristics of the poverty population. The factors will also be used in the subsequent regression analysis.

The use of Principal Components Analysis, and the closely related technique of Common Factor Analysis (Flury 1988), to investigate the inherent dimensions within socioeconomic data is well documented. Thompson et al. (1962) used the method to identify latent dimensions within a data set which contained nine variables associated with economic health. Within poverty research, factor analysis has been used in a previous study by Brunn and Wheeler (1971). In research aimed at identifying spatial dimensions of U.S. poverty, factors identified were seen to indicate that poverty was associated with several conditions.

Twenty five variables, as listed in Table 2, were used in the Principal Components Analysis. Data for all 3109 United States counties, or county equivalents, are included in the analysis. A Varimax rotation was conducted to more clearly illustrate the dimensions within the data and to facilitate factor interpretation. The criteria used for the retention of a factor was a minimum eigenvalue of one. Interpretation

Table 3. Pearson Product Moment Correlation Matrix.

VARIABLE	INDEX	POPCH	URBAN	WHITE	BLACK
INDEX	1.000				
POPCH	.152 •	1.000			
URBAN	.446 •	.020	1.000		
WHITE	.350 •	.073 •	-.094 •	1.000	
BLACK	-.318 •	-.097 •	.059 •	-.899 •	1.000
INDIAN	-.145 •	.043 *	-.070 •	-.270 •	-.089 •
SPANISH	-.090 •	.031	.173 •	-.125 •	-.098 •
YOUNG	-.147 •	.072 •	-.042 *	-.197 •	.052 •
TEENS	-.318 •	-.296 •	-.313 •	.184 •	-.113 •
OLD	-.223 •	.055 •	-.168 •	-.331 •	.234 •
PERFEM	-.238 •	-.090 •	.348 •	-.787 •	.764 •
TEENBIR	-.486 •	-.048 •	-.108 •	-.469 •	.457 •
PHYSIC	.302 •	.007	.445 •	-.035	.026
HOSP	.082 •	-.136 •	.205 •	-.004	.005
PLUMB	-.585 •	-.038 *	-.415 •	-.433 •	.432 •
OVERCR	-.493 •	.095 •	-.088 •	-.662 •	.435 •
HSCHOOL	.666 •	.155 •	.344 •	.412 •	-.445 •
COLLEGE	.506 •	.217 •	.476 •	.052 •	-.094 •
UNEMPL	-.262 •	.086 •	-.102 •	-.070 •	.049 •
COMMUT	.100 •	.257 •	-.190 •	-.079 •	.162 •
LABFEM	.069 •	.058 •	.443 •	-.345 •	.395 •
LABMAN	.028	-.046 *	.016	-.115 •	.256 •
LABRET	.253 •	.147 •	.450 •	.166 •	-.154 •
LABPROF	.027	-.105 •	.313 •	-.065 •	-.009
DEPEND	-.524 •	-.247 •	-.448 •	-.091 •	.053 •
LABPART	.668 •	.039 *	.307 •	.239 •	-.120 •

Table 3. Continued

VARIABLE	INDIAN	SPANISH	YOUNG	TEENS	OLD
INDIAN	1.000				
SPANISH	.003	1.000			
YOUNG	.286 •	.213 •	1.000		
TEENS	-.112 •	-.112 •	-.414 •	1.000	
OLD	.236 •	.192 •	.648 •	-.452 •	1.000
PERFEM	.135 •	.039 *	-.011	-.245 •	.188 •
TEENBIR	.068 •	.080 •	.097 •	.047 •	.285 •
PHYSIC	-.043 *	.008	-.197 •	-.167 •	-.275 •
HOSP	-.002	-.034	-.117 •	.079 •	-.204 •
PLUMB	.136 •	-.017	.070 •	.013	.323 •
OVERCR	.428 •	.484 •	.509 •	-.256 •	.586 •
HSCHOOL	.012	-.086 •	.009	-.207 •	-.320 •
COLLEGE	-.008	.057 •	-.133 •	-.365 •	-.341 •
UNEMPL	.122 •	-.060 •	.038 *	-.131 •	.210 •
COMMUT	-.106 •	-.142 •	-.093 •	-.204 •	.245 •
LABFEM	-.018	-.157 •	-.319 •	-.123 •	-.213 •
LABMAN	-.157 •	-.272 •	-.175 •	-.134 •	.171 •
LABRET	-.124 •	.072 •	-.058 •	-.007	-.252 •
LABPROF	.151 •	.029	-.167 •	-.043 *	-.292 •
DEPEND	.135 •	.082 •	.333 •	.603 •	.400 •
LABPART	-.141 •	-.078 •	-.220 •	-.182 •	-.322 •

Table 3. Continued

VARIABLE	PERFEM	TEENBIR	PHYSIC	HOSP	PLUMB
PERFEM	1.000				
TEENBIR	.443 •	1.000			
PHYSIC	.259 •	-.193 •	1.000		
HOSP	.105 •	-.028	.270 •	1.000	
PLUMB	.356 •	.461 •	-.218 •	-.138 •	1.000
OVERCR	.457 •	.469 •	-.147 •	-.101 •	.537 •
HSCHOOL	-.347 •	-.703 •	.325 •	.095 •	-.664 •
COLLEGE	.058 •	-.468 •	.523 •	.143 •	-.366 •
UNEMPL	.254 •	.225 •	-.042 *	-.090 •	.347 •
COMMUT	.042 *	.087 •	-.204 •	-.195 •	.233 •
LABFEM	.562 •	.061 •	.357 •	.187 •	-.026
LABMAN	.284 •	.264 •	-.054 •	-.112 •	.215 •
LABRET	-.041 *	-.256 •	.235 •	.071 •	-.414 •
LABPROF	.206 •	-.195 •	.415 •	.386 •	-.112 •
DEPEND	-.137 •	.267 •	-.418 •	-.090 •	.246 •
LABPART	-.178 •	-.478 •	.300 •	.085 •	-.473 •

Table 3. Continued

VARIABLE OVERCR HSCHOOL COLLEGE UNEMPL COMMUT

	OVERCR	HSCHOOL	COLLEGE	UNEMPL	COMMUT
OVERCR	1.000				
HSCHOOL	-.485 •	1.000			
COLLEGE	-.199 •	.692 •	1.000		
UNEMPL	.192 •	-.202 •	-.224 •	1.000	
COMMUT	.011	-.207 •	-.127 •	.060 •	1.000
LABFEM	-.066 •	.022	.297 •	.059 •	.035
LABMAN	-.056 •	-.354 •	-.303 •	.239 •	.315 •
LABRET	-.248 •	.413 •	.312 •	-.037 *	-.240 •
LABPROF	-.052 •	.294 •	.511 •	.066 •	-.196 •
DEPEND	.288 •	-.418 •	-.646 •	.013	-.085 •
LABPART	-.431 •	.508 •	.520 •	-.598 •	-.027

VARIABLE LABFEM LABMAN LABRET LABPROF DEPEND

	LABFEM	LABMAN	LABRET	LABPROF	DEPEND
LABFEM	1.000				
LABMAN	.383 •	1.000			
LABRET	.220 •	-.319 •	1.000		
LABPROF	.328 •	-.297 •	.185 •	1.000	
DEPEND	-.375 •	-.088 •	-.189 •	-.291 •	1.000
LABPART	.209 •	.007	.141 •	-.016	-.473 •

•   Significant at the .01 level
*   Significant at the .05 level

of the factors generated is based on the loading of each variable with respect to the factor. The factors generated are used in the next step of the analysis which seeks to model the concomitants of income poverty and the characteristics of the poverty population for the nation as a whole.

3. A regression model explores the relationship between the income index and the concomitants of income poverty and characteristics of the poverty population identified in step 2. The dependent variable used is the income index (INDEX). The natural log of INDEX is used in the regressions for two reasons. Conceptually it can be argued that at low levels of adjusted income, any change in available income can be expected to result in a relatively large changes in the independent variables. A poor person is likely to use disposable income to reduce deficits in some concomitants of income poverty. As an individual becomes more affluent, a unit change in INDEX is likely to result in a smaller change in some independent variables that are not fixed. A more affluent person will likely have already purchased most of the desired services within the areas of housing, education, and health that will be reflected in a change in the independent variables. Methodologically, using the logarithm considerably reduces the large variation in the index. The log transformation results in a more powerful model in which all terms are significant, which is not the case in models without log transformation. The use of a log transformation for income data is commonly used in the literature (Cowell 1977; Oster, Lake, and Oksman 1978).

The independent variables are given by the factors generated by the Principal Components Analysis. Use of these factors within multiple regression essentially eliminates the problem of multicollinearity, without eliminating variables arbitrarily. This step of the analysis uses all counties in the contiguous United States to define a general model of the concomitants of income poverty and the characteristics of the poverty population for the nation as a whole. The regression model is given by Equation 5.1.

$$LogINDEX = k_0 + k_1 \ FACTOR1 + k_2 \ FACTOR2 \ ...+ k_n \ FACTORn$$

(Equation 5.1)

$k_0, k_1, k_2 \ ... \ k_n$ represent the parameters of the equation and express the direction and strength of each poverty dimension on the income index.

The regression was conducted using SAS version 5.18. The model was defined using the "Reg" procedure and residuals were checked for compliance with the procedure's assumptions using the "Plot" procedure. The residuals were found to largely comply with the procedure's assumptions.

Results and Discussion

The analysis of research question 2 involved three phases: the correlation analysis, the Principal Component Analysis, and regression modeling.

Correlation Analysis. The results of the correlation analysis are presented in Table 3. Not surprisingly, of the 325 relationships represented in the matrix, 279 show significant correlations at the .01 level and an additional 13 are significant at the .05 level. Overall the analysis supports the use later of Principal Components Analysis to group highly correlated variables into factors or dimensions.

Table 3 shows that the relationships between the dependent variable INDEX and the independent variables are in the expected directions discussed previously. Eleven of the twenty five independent variables (POPCH, URBAN, WHITE, PHYSIC, HOSP, HSCHOOL, COLLEGE, COMMUT, LABFEM, LABRET, and LABPART) are significantly and positively related to the dependent variable INDEX, while twelve variables (BLACK, INDIAN, SPANISH, YOUNG, TEENS, OLD, PERFEM, TEENBIR, PLUMB, OVERCR, UNEMPL, and DEPEND) are significantly negatively related to INDEX. Two variables, LABMAN and LABPROF, show no significant correlation with INDEX. It appears that the percentage of the labor force employed in manufacturing and in professional and related services are not significantly related to adjusted per capita income. This lack of a clear relationship with index may be a product of variability within the manufacturing and professional employment categories. While some jobs such as in top management may be high payed, many others will be low payed and may be minimum wage.

Principal Components Analysis. Twelve factors meeting the eigenvalue criterion emerged from the Principal Components Analysis. The twelve factors together explained almost 90 percent of the information contained within the original data set (Table 4). While the rotation results in factor loadings with values that are more toward the extremes of |1| and 0 and hence are more easily interpretable, the communality of the rotated factors remains unchanged. Thus, the twelve factors used in this study lose little of the information contained in the original data set, and can be confidently substituted for the original variables in later analyses.

Not only is most of the information expressed by the original independent variables retained, but also all twenty five variables are well represented by the twelve factors generated. Table 5 displays the communality estimates for each original variable. Communality values indicate the amount of original information expressed by a variable that is retained in the factors generated. A value of 1 indicates all original information in a particular variable is contained within the factors generated. Communality values for the twenty five independent variables vary between a high of 0.968 (TEENS) to a low of 0.769 (PLUMB). Eleven variables have over 90 percent of their original information contained within the factors generated. Only two variables, TEENBIR and PLUMB, have communalities less than 0.8 and even these two variables have over three quarters of their original information contained within the factors (Table 5).

The variables captured by each of the twelve factors are revealed by the factor loading matrix. Each variable has a factor loading between +1 and -1 on each factor. Factor loadings close to the extreme values indicate a close relationship between the original variable and that factor. Table 6 presents the variables that were considered significantly loaded on each factor, and a tentative interpretation of each factor.

Factor 1 is interpreted to represent the black-white duality or racial division. The factor is strongly and positively associated with the percent of African American population, and strongly negatively associated with the percent of white population. In addition to BLACK, factor one is also strongly positively linked to PERFEM, the percentage of female headed families. The propensity for a family to be female headed has been observed to be higher among African American families than among white families. Thus, high loadings on PERFEM are consistent with the interpretation of Factor 1 offered. Factor 1 is anticipated to be positively associated with poverty and thus negatively associated with the dependent variable INDEX.

Factor 2 is loaded positively by TEENS and DEPEND, and negatively by COLLEGE and URBAN. A high value on Factor 2 means the population of a county has fewer members between the ages

Table 4. Eigenvalues of the 1980 PCA factors.

Factor	Rotated Eigenvalues	% Varience Explained
1	3.94	15.76
2	2.56	10.24
3	2.42	9.68
4	2.22	8.88
5	1.97	7.88
6	1.71	6.84
7	1.51	6.04
8	1.38	5.52
9	1.22	4.88
10	1.21	4.84
11	1.10	4.40
12	1.10	4.40
		89.36

Table 5. Communality Estimates for the Variables Used in
the 1980 Principal Components Analysis.

Variable	Communality
POPCH	.933
URBAN	.816
WHITE	.944
BLACK	.966
INDIAN	.954
SPANISH	.947
YOUNG	.876
TEENS	.968
OLD	.897
PERFEM	.892
TEENBIR	.789
PHYSIC	.900
HOSP	.940
PLUMB	.769
OVERCR	.924
HSCHOOL	.874
COLLEGE	.874
UNEMPL	.900
COMMUT	.896
LABFEM	.830
LABMAN	.895
LABRET	.831
LABPROF	.846
DEPEND	.962
LABPART	.880

Table 6. Significant Factor Loadings and Interpreted
Factor Meanings, 1980 Factors.

Factor	Loading		Meaning
1	BLACK	.947	Black - white duality.
	PERFEM	.858	
	WHITE	-.935	
2	DEPEND	.896	Non-Yuppieness.
	TEENS	.794	
	COLLEGE	-.618	
	URBAN	-.407	
3	YOUNG	.896	Extreme age population.
	OLD	.824	
	TEENS	-.520	
4	LABRET	.833	Urban retail economy.
	URBAN	.674	
	PLUMB	-.596	
5	UNEMPL	.919	Employment status.
	LABPART	-.770	
6	LABMAN	.869	Manufacturing/blue collar economy.
	LABPROF	-.442	
7	PHYSIC	.850	Health service economy.
8	SPANISH	.948	Spanish culture and ethnicity.
	OVERCR	.477	
9	INDIAN	.941	Native American culture and ethnicity.
10	HOSP	.949	Health care availability.
11	POPCH	.916	Growth.
12	COMMUT	.858	Labor mobility.

of 18 and 65 years and is unlikely to be highly educated or urbanized. This description seems to be the opposite of a population that is commonly described as "Yuppie." Factor 2 is therefore interpreted to represent "non-yuppieness." The relationship between factor two and the income index is expected to be negative.

Factor 3 is a variable that is associated with age structure. Strong positive loadings can be seen at both extremes of age. This factor is assessed to represent extreme aged population, that is a population that has high proportions of the old or very young. Factor 3 is interpreted to represent extreme age population. Factor 3 is expected to show a positive relationship to poverty.

The highest positive loading associated with Factor 4 is LABRET. This factor also has high positive loading on URBAN. The highest negative loading associated with Factor 4 is PLUMB. Since retail employment is concentrated in urban areas this factor can be interpreted to represent the size of a county's urban retail economy. The high negative loading on PLUMB is consistent with this interpretation, since lack of plumbing facilities is overwhelmingly a rural problem. It is a clear expectation that Factor 4 will be positively linked to income, and thus negatively associated with poverty.

Factor 5 is loaded heavily both positively and negatively by variables that represent employment status. UNEMPL is positively loaded, while LABPART is negatively loaded. This factor clearly can be interpreted to represent employment status. Factor 5 is anticipated to have a positive relationship to poverty.

High proportions of the workforce employed within manufacturing (LABMAN) are associated positively with Factor 6. The highest negative loading is with the percentage of the workforce employed in health, education, and other professional and related services (LABPROF). Thus, this factor appears to represent a manufacturing or blue-collar economy. The literature suggests that employment in manufacturing provides relatively high wages to workers and economic prosperity to a region. Therefore, an economy characterized by high levels of manufacturing activity is expected to be associated with high income and low poverty levels.

Factor 7 has high positive loadings on two variables, PHYSIC and COLLEGE. This factor appears to represent a health service economy. This interpretation is consistent with the two variables that contribute most to Factor 7. Health service economies are anticipated to be associated with relatively high income.

Factor 8 has high loadings on SPANISH and OVERCR. It seems this factor represents Spanish ethnicity and culture. High levels of poverty and relatively low income are expected to be experienced in counties with high scores on Factor 8. Factors 9 through 12 each have high positive loadings on only one variable and no high negative factor loadings. Thus, it seems that these factors represent essentially the same information as an original variable. The expected direction of the relationship between these variables and adjusted income have been discussed previously.

Factor 9 is strongly loaded by INDIAN and, therefore, Factor 9 represents Native American culture and ethnicity. HOSP is the original variable that contributes most information to Factor 10. Factor 10 is interpreted to indicate levels of health care availability. Factor 11 is imbued with its meaning by the original variable POPCH. Since POPCH is the percent population change from 1970 to 1980, and the loading is positive, Factor 11 represents growth at a county level. The final factor, Factor 12, has a strong loading by the original variable COMMUT. Since COMMUT is the percentage workers who work outside the county of residence, Factor 12 is said to represent labor mobility. The concomitants of income poverty are illustrated by the factors that were obtained by conducting the principle components analysis. Each factor represents a dimension of poverty or a characteristic of the poverty population that can be used to model poverty in the United States.

Regression Analysis. The results of the regression given by Equation 5.1 are provided in Appendix B1. The results show that all parameters are significant at the .01 level. The direction of the relationship between LogINDEX and each of the factors was as anticipated. High income was indeed negatively related to blackness, non-yuppieness, extreme age population, unemployment, Spanish ethnicity, and Native American ethnicity, and positively related to the urban retail economy, the health service economy, the manufacturing/blue collar economy, health care availability, growth, and labor mobility.

Since the factors are standardized variables, the magnitudes of the parameter estimates can be directly compared to assess the degree of association of each factor with the income index. Four Factors stand out as contributing substantially to explanation of the magnitude of the dependent variable. The four most important factors associated with available income are the non-yuppieness factor (Factor 2), the factor which expresses the black-white duality (Factor 1), the unemployment factor (Factor 5), and the factor which gives the extent of the urban retail economy (Factor 4). Thus, a county with a high percentage of black population, with a high percentage of young or old people, whose population lacks higher education, which has low levels of urbanization and employment, and which lacks an urban retail economy is most likely to suffer from low levels of adjusted per capita income.

A somewhat less substantial influence on the dependent variable is exerted by three factors. The factors representing manufacturing and health service economies (Factor 6 and Factor 7) have a positive impact on adjusted per capita income. In contrast, counties with large numbers of people of Spanish ethnicity (Factor 8) are negatively associated with adjusted per capita income.

A third level of influence by the identified factors on the dependent variable LogINDEX is exerted by two factors. The Native American culture and ethnicity factor (Factor 9), is negatively associated with adjusted per capita income. In contrast, the labor mobility factor (Factor 12) has a positive association with the income index.

At a fourth level, three factors exert a small, but significant influence on adjusted per capita income. The factor expressing relatively large numbers of the old or very young (Factor 3) is negatively related to adjusted per capita income. Finally, the factors expressing health care availability and growth (Factors 10 and 11) are associated with somewhat elevated values of adjusted per capita income.

The twelve factors combine to explain 70 percent of the variance in the income index. The F probability of .0001 indicates that the model is significant under rigorous significance criteria. It appears that the model presented is relatively powerful in explaining the magnitude of adjusted per capita income at the national level. These results obtained from the spatially invariant national model of poverty provide a view of the characteristics of the poverty population and of the concomitants of income poverty.

The results of the model of poverty indicate that race and ethnicity are important characteristics of the poverty population. Non-white, or minority Americans, are more likely to experience poverty than their white counterparts. African Americans, Native Americans, and Americans who are of Spanish origin all make up a disproportionate portion of the poverty population. It is this racial and ethnic dimension that is the most striking characteristic of the population that struggle to share the affluence of the nation. This conclusion concerning the importance of race and ethnicity comes as no surprise. While many researchers have focused on the racial dimension of urban poverty, many have also pointed to the link between rural poverty and race (Dhillon and Howie 1986; McCormick 1988). Most research links poverty with the African American population and this research confirms this link. However, there is limited discussion of the link between poverty and other minority populations. This research clearly indicates that both Native Americans and those of Spanish ancestry are disproportionately economically disadvantaged. These groups appear to be part of the invisible poor whose plight is seldom brought into the public or academic spotlight.

The racial concomitant of income poverty which places many African Americans amongst the poor, is part of perhaps the most complex association of characteristics. The African American racial dimension cannot be separated from the high incidence of female headed households and teen births, or from overcrowded and substandard housing. This racial dimension is also linked to a lack of education as evidenced by high school dropout rates. All these characteristics seem reminiscent of what has been called the urban underclass. The urban underclass is invariably described as non-white and uneducated, as outside the economic mainstream, and as disproportionally composed of female headed households and teenaged mothers. It seems acceptance of a distinctly urban underclass may be questionable. This research points to the existence of a very similar group that live in geographic isolation in rural America. The geographic location of this rural counterpart of the urban underclass will be discussed in Chapter 6, but what seems clear is that a more proper term if one wants to label this group may be the "non-white underclass." However, although individualistic underclass explanations of the underclass can be extended to include the rural black poor, structural explanations cannot. Structural explanations that revolve around the shift from manufacturing to services and the changing location of jobs have little applicability in the rural south. Other possible structural factors must be explored; this is done on a regional basis in Chapter 6.

A second important characteristic of the poverty population involves its demographic structure. The young and the elderly are over represented in the poverty population. This finding is in part consistent with previous research. Much has been written about the tragedy of the nation's poor children (Goldstein 1986; O'Hare 1988). Undoubtedly, children are often the victims of poverty. However, it seems that the elderly are still a group that suffer inordinately from poverty, as evidenced by the link between the dependency rate and impoverishment. This is despite the assertion by some that the war on poverty with respect to America's older citizens has been won. The social spending that is concentrated on the elderly has not extracted this group from poverty. This research supports the contention of Ruggles (1992) that old age often remains an experience that lacks material comfort and financial security, and that many of the elderly exist on incomes just above the poverty line and so are not officially categorized as poor. Thus, the elderly as a group cannot be discounted. However, the cores of poverty are more associated with concentrations of young children than with the elderly. Young children that make up portions of the poverty population must be part of family units that contain poor young adults. Thus, the profile of the poor in poverty cores, with the exception of Appalachia, is of minority young adults with children.

A third dimension that can be considered as characteristic of the poverty population concerns employment. Employment status and mobility associated with employment are both of importance. While much has been written about the growing numbers of working poor in America, this research suggests that employment is still generally associated with non-poverty. Conversely the evidence presented here suggests that unemployment is clearly linked to poverty. Thus, on a general level lack of employment, especially for minorities, was a serious problem in 1980.

Lack of mobility is also an important characteristic of the poverty population. It is clear that mobility is not merely a result of personal characteristics, but as with other such characteristics is a product of interaction with structural factors. The transportation network in a region is an obvious prerequisite for mobility. Also, to be mobile people must command enough financial resources to own and operate a vehicle or to use other available transport. Thus, mobility can be envisaged to be enabled by some threshold of income, and poverty and immobility to be locked together in a cycle that is difficult to break.

The aforementioned characteristics of the poverty population interact with several structural concomitants of income poverty. Some general conclusions are clear. It is the structure of the economy that dominates those factors external to the individual. County economies that are focused on manufacturing and health service activity offer residents higher income than counties that lack such a focus. In addition, urban areas where retail activity is concentrated also provide relatively high income for workers. The influence of retail, health service, and manufacturing activity can be explained by the fact that all are major providers of jobs. Again, it seems clear that despite all that has been written concerning the problems of urban areas and the rise of the low income service sector, it is the existence of an urban retail economy that is the most influential structural factor raising income.

Of the three economic structural factors identified, manufacturing, the traditional provider of relatively high income, is the least powerful influence. Many communities still seem to work to attract manufacturing to their county; a recent example is the intense competition to attract the Mercedes car plant that was won by Tuscaloosa, Alabama. This research indicates that winning the competition for such industry may not be the best route to economic prosperity. Rather, promotion of urban retail centers is revealed to be a better avenue to pursue in the search for an improved local economy. The shift from manufacturing to services does not emerge as a major factor that explains poverty at the county level. This again indicates that structural explanations applied in the analysis of the urban underclass are inadequate to explain the spatial concentration of poverty outside urban areas. A separate view of each poverty region is indicated. While large metropolitan areas can be said to share somewhat similar characteristics and hence may appropriately be brought under a single theoretical umbrella, the same is not true of the five poverty cores identified by this research.

In addition to economic structure, lack of health care availability and population growth are linked to income poverty. It is not surprising that lack of health care availability goes hand in hand with lack of income. The United States is a nation with a limited public health care system where the ability to pay for health care is fundamental to its availability. While one might expect education to emerge as a clear concomitant of income poverty, it does not. There may be questions concerning the quality of education, but unlike health care, education is available to all. It is not only available but is required. However, a county may graduate a large percentage of its children from high school but this will not help the poverty situation if jobs are not available or if high school graduation is no indication of real educational attainment. Without other changes there seems to be no indication in this research that increasing rates of college and high school graduation will alleviate poverty. In contrast, this research clearly indicates support for a universal health care system that covers all Americans. While decent housing and a quality education are issues of concern, they do not emerge as clear and distinct dimensions of poverty as does lack of health care. This finding of the importance of health care availability is particularly germane at a time when fundamental health care reform is a prominent political topic.

The connection between population growth and relative affluence may be a product of people moving toward areas of economic prosperity. Counties where the economy is strong, the urban retail economy is growing, and jobs are available are likely to attract a growing number of residents. It is unlikely that population growth per se brings prosperity. On the contrary, population growth without a growing economic structure to support it will probably result in increased poverty levels. Thus, it seems reasonable to suggest that results of population growth are likely to vary with other conditions such as the availability of jobs, and the prevailing economic structure and context.

There are clearly identifiable concomitants of income poverty and characteristics of the poverty population, and the conclusions drawn above present a general picture for the United States. However, at least two critical questions remain to be answered. First, are these concomitants and characteristics spatially invariant or are there any significant geographic variations that can be delineated? If such spatial variations are found to exist the conclusions drawn by analysis of the national model may not be valid for all geographic areas. In addition is it possible that factors that are relatively insignificant nationally become important in some geographic areas, or conversely those important at the national level become less prominent in certain regions? Second, has the geography of poverty remained spatially stable in recent years, especially over the 1980s with the extensive economic, political, and social changes that characterized the decade. Another related question concerning temporal changes involves possible changes in the nature of poverty? While poverty may have remained spatially stable during the 1980s it is possible that the nature of poverty being experienced in poverty cores might have undergone a significant transition.

NOTES

1. Sources:
   a. 1980 data. 1980 United States Census. Reported in U.S. Department of Commerce. 1983. County and City Data Book. U.S.Bureau of the Census.
   b. 1990 data. 1990 United States Census. Reported on Summary

VI

A Spatially and Temporally
Varying Model of Poverty

INTRODUCTION

The previous chapter explored the spatial distribution of poverty in the United States and identified five major cores of poverty. A general model of poverty for the nation as a whole was developed. The model illustrated the relative importance of the various concomitants of income poverty and characteristics of the poverty population.

This chapter looks at spatial variation in the general model of poverty. The focus is on variation between urban and rural counties and also spatial variation by region. The national spatially invariant model may mask important spatial variations in the importance of identified dimensions of poverty.

This section also tests the temporal stability of the general model of poverty from 1980 to 1990. Changes may have occurred in the importance of the various dimensions associated with income poverty. Dimensions strongly associated with poverty in 1980 may have either strengthened or weakened during the the 1980s.

SPATIAL VARIATION IN THE CONCOMITANTS OF POVERTY AND CHARACTERISTICS OF THE POVERTY POPULATION

Data and Methodology

The factors generated by the Principal Components Analysis are used in two ways to explore the spatial variation in the concomitants of income poverty and characteristics of the poverty population. First, the factors are mapped in Figures 7 to 18. These maps visually display the spatial variation at the national level of the dimensions identified within the original data set by the Principal Components Analysis.

The second step of the analysis seeks to explore spatial variation in the nature of poverty. This is accomplished by looking for spatial variation in the general model of poverty (Equation 5.1) by use of the Expansion Method (Casetti 1972). The general model is first "expanded" to explore urban versus rural distinctions in poverty. This seems appropriate since the existing literature consistently draws the distinction between urban and rural poverty, rather than viewing urbanness as a continuum. Much has been written about "urban" poverty as separate and distinct from "rural" poverty. In this study any county with a population that was less than 50 percent urban (as defined in the U.S. Census) was categorized as rural. All other counties were classified as urban (Figure 19). Secondly, spatial variation in poverty by region is investigated by expansion of the spatially invariant national model. The regionalization selected is that of the geographic division as defined by the Bureau of the Census (Note: Alaska and Hawaii were excluded from the "Pacific" division in this study). There are nine geographic divisions, composed of aggregations of states, that are widely used by the Bureau of the Census (Figure 20). These census geographic divisions were selected as the preferred regionalization for two reasons. First, these geographic divisions have been used by the Bureau of the Census since 1910. Thus, their use allows comparison of results with a majority of available information and research. Second, these divisions were defined so as to be relatively homogeneous in terms of physical characteristics, characteristics of the population, and social and economic characteristics (Kaplan and Van Valey 1980). The definition of these geographic divisions under the principle of demographic and socioeconomic homogeneity would appear to make them an appropriate choice in a study of poverty. It must be pointed out that these geographic divisions are not without problems. Jackson et al. (1983) point out that the "South Atlantic" division spans a large area north to south that is difficult to view as homogeneous. The same problem exists within the large "Mountain" region (Jackson et al. 1983). Also, state boundaries may not represent ideal boundaries between regions, but they do offer continuity over time (Kaplan and Van Valey 1980).

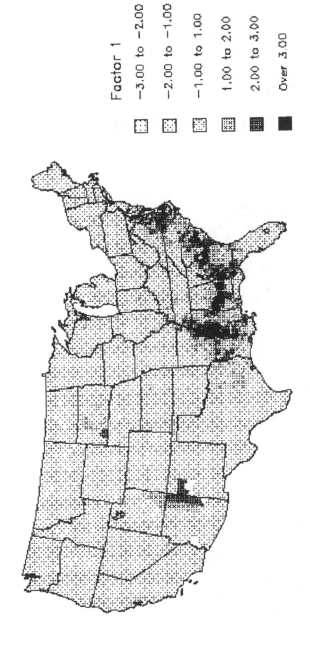

Figure 7. Spatial Variation in the Black-White Duality in the U.S. - 1980

Factor 1

☐ -3.00 to -2.00

☐ -2.00 to -1.00

☐ -1.00 to 1.00

☒ 1.00 to 2.00

☐ 2.00 to 3.00

■ Over 3.00

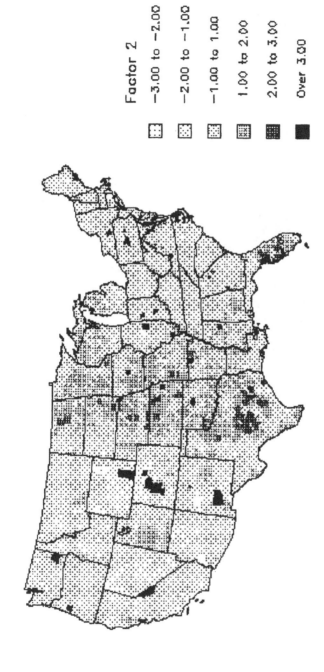

Figure 8. Spatial Variation in Non-yuppieness in the U.S. - 1980

Factor 2

▦ −3.00 to −2.00

▦ −2.00 to −1.00

▦ −1.00 to 1.00

▦ 1.00 to 2.00

▦ 2.00 to 3.00

■ Over 3.00

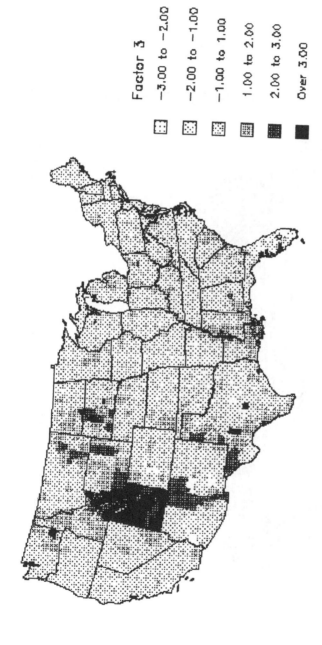

Figure 9. Spatial Variation in Extreme Age Population in the U.S. - 1980

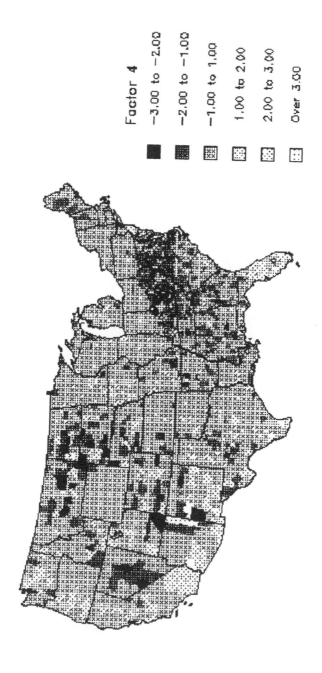

Factor 4

■ −3.00 to −2.00

▦ −2.00 to −1.00

▧ −1.00 to 1.00

▨ 1.00 to 2.00

▩ 2.00 to 3.00

▫ Over 3.00

Figure 10. Spatial Variation of the Urban Retail Economy in the U.S. - 1980

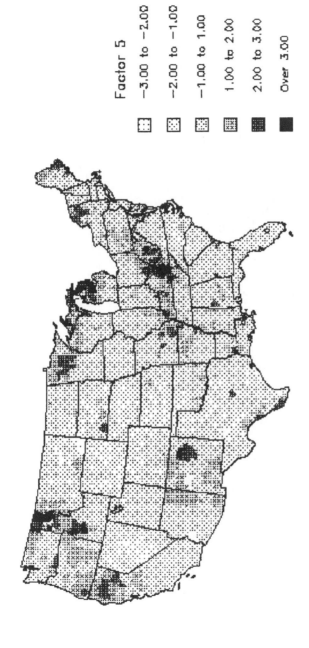

Figure 11. Spatial Variation in Employment Status in the U.S. - 1980

Factor 5

▦	−3.00 to −2.00
▦	−2.00 to −1.00
▦	−1.00 to 1.00
▦	1.00 to 2.00
▦	2.00 to 3.00
■	Over 3.00

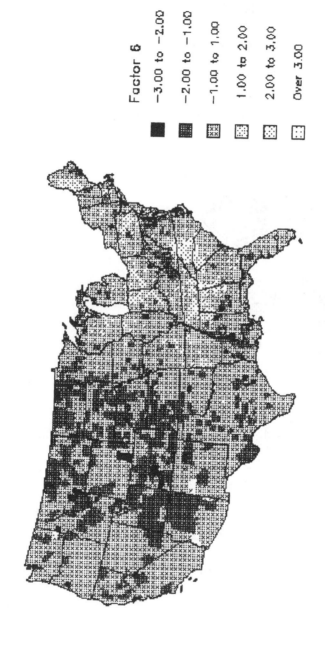

Figure 12. Spatial Variation of the Manufacturing/Blue Collar Economy in the U.S. - 1980

114

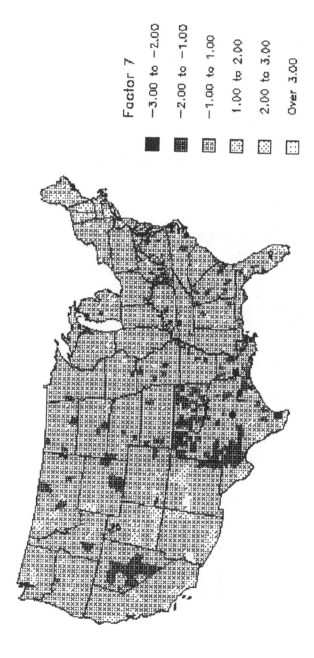

Figure 13. Spatial Variation of the Health Service Economy in the U.S. - 1980

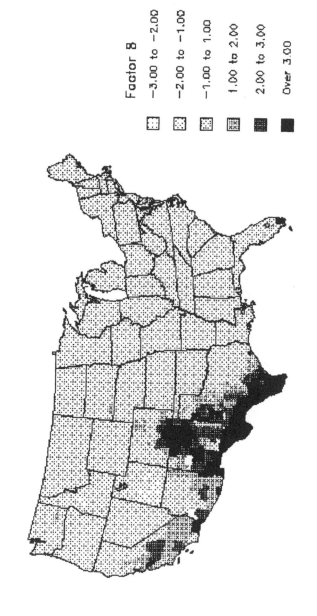

Figure 14. Spatial Variation of Spanish Culture and Ethnicity in the U.S. - 1980

Factor 8

⬚ -3.00 to -2.00
⬚ -2.00 to -1.00
⬚ -1.00 to 1.00
▦ 1.00 to 2.00
▩ 2.00 to 3.00
■ Over 3.00

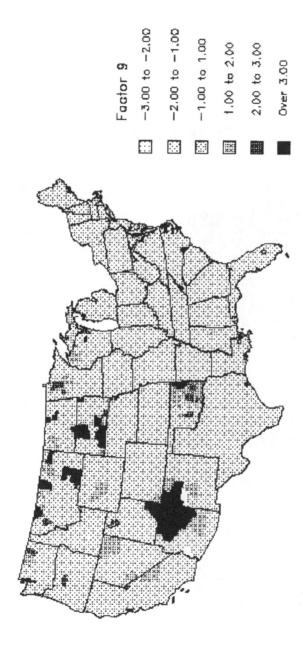

Figure 15. Spatial Variation in Native American Culture and Ethnicity in the U.S. - 1980

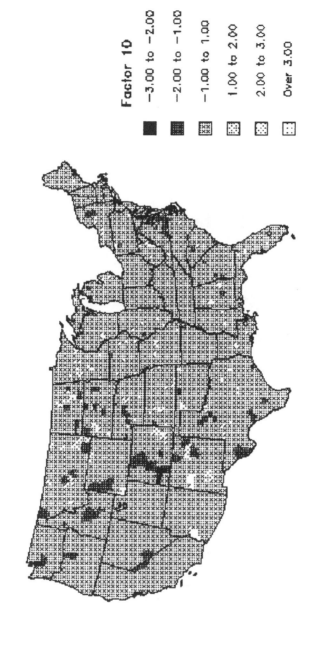

Figure 16. Spatial Variation of Health Care Availability in the U.S. - 1980

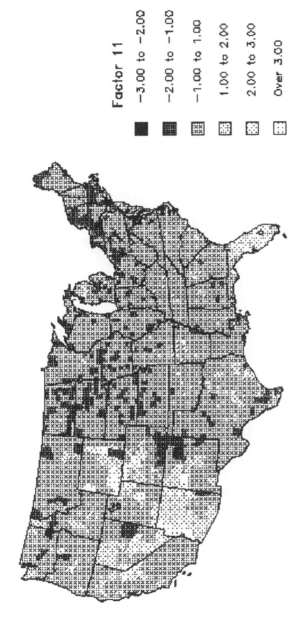

Figure 17. Spatial Variation of Population Growth in the U.S. - 1980

Factor 11

■ -3.00 to -2.00
■ -2.00 to -1.00
▨ -1.00 to 1.00
▨ 1.00 to 2.00
▨ 2.00 to 3.00
▨ Over 3.00

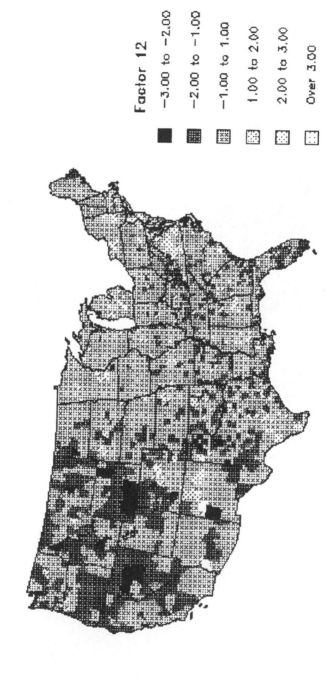

Figure 18. Spatial Variation of Labor Mobility in the U.S. - 1980

Factor 12

- −3.00 to −2.00
- −2.00 to −1.00
- −1.00 to 1.00
- 1.00 to 2.00
- 2.00 to 3.00
- Over 3.00

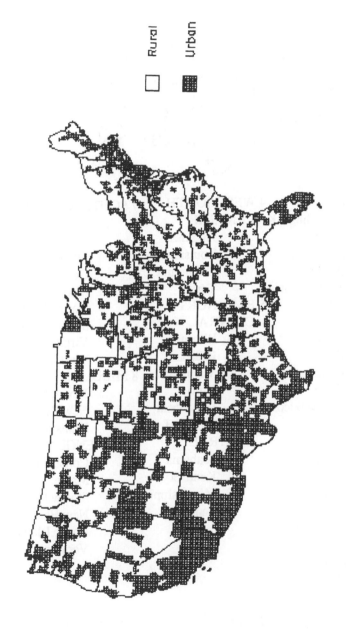

Rural ☐

Urban ▧

Figure 19. The Distribution of Urban and Rural Counties in the U.S. - 1980

1 PACIFIC	4 WEST SOUTH	7 SOUTH ATLANTIC
(excluding AK	CENTRAL	Delaware
& HA)	Arkansas	D.C.
California	Louisiana	Florida
Oregon	Oklahoma	Georgia
Washington	Texas	Maryland
2 MOUNTAIN	5 EAST NORTH	North Carolina
Arizona	CENTRAL	South Carolina
Colorado	Illinois	Virginia
Idaho	Indiana	West Virginia
Montana	Michigan	8 MIDDLE ATLANTIC
Nevada	Ohio	New Jersey
New Mexico	Wisconsin	New York
Utah	6 EAST SOUTH	Pennsylvania
Wyoming	CENTRAL	9 NEW ENGLAND
3 WEST NORTH	Alabama	Connecticut
CENTRAL	Kentucky	Maine
Iowa	Mississippi	Massachusetts
Kansas	Tennessee	New Hampshire
Minnesota		Rhode Island
Missouri		Vermont
Nebraska		
North Dakota		
South Dakota		

Figure 20. Geographic Divisions of the Conterminous United States.

Urban-Rural Variation. To investigate urban - rural variation of poverty a dummy variable is created to identify whether a county is urban or rural. Thus, when a county has a population that is 50 percent or more urban the value of the variable TYPE is said to be 1. Using the criteria of at least 50 percent urban population, 975 counties are designated as urban. Counties that are less than 50 percent urban have the value of TYPE set to 0. The number of counties with a population that is less than 50 percent urban is 2134 (Figure 19).

The urban/rural variable is used within an expansion of the regression model used to represent poverty at the national level (Equation 5.1). The effect of urban/rural context on the model of poverty was developed using the Expansion Method (Casetti 1972). The Expansion Method allows investigation of contextual variation in model parameters. The Expansion Method has been widely used to explore both temporal and spatial variation in general models of socioeconomic relationships. Pandit and Casetti (1989) used the method to illustrate temporal shifts in a general model of development, while Casetti and Jones (1983) investigated regional variation in a general model of manufacturing productivity by application of the Expansion Method. In a study related to poverty, Jones (1984) effectively utilized the Expansion Method to illustrate spatial variation in AFDC participation and showed that participation varied spatially in response to such things as unemployment.

Several points can be strongly made in support of the use of the Expansion Method in a geographic study such as is presented in this research. Any assumption of spatial stability of parameters in a general model is suspect within a geographic study, and the Expansion Method provides a technique for assessing spatial variation in a general relationship (Casetti 1972; Jones 1984). The Expansion Method allows for modeling of relationships in a real world context. Assumptions of independence from geographic and historical context, which underlie conventional models, do not provide a realistic perspective of the phenomenon being studied. Further, such models deny spatial variation which is fundamental to geography (Foster 1991).

Using the Expansion Method involves three primary steps: 1. modeling of the initial relationship to be investigated for contextual variation, 2. the modeling of variation in the relationship using expansion equations. The expansion equations relate the initial model parameters to the contextual variables, 3. the expansion variables are substituted in the initial model to produce a terminal model, which allows contextual variation.

The relationship between income poverty (INDEX) and the concomitants of poverty and characteristics of the poverty population, was represented by the model presented previously (Equation 5.1). This is the "initial model" on which the expansion was performed. This initial model represents the contextually invariant relationship between INDEX and the twelve factors generated by the Principal Components Analysis from the twenty five selected independent variables.

The differential effect of the twelve factors on INDEX, by urban versus rural character, was investigated using a series of expansion equations. The parameters in the initial model were expanded by the variable TYPE as follows:

$$k_i = k_{i0} + k_{i1} \text{TYPE}$$

(Equation 6.1)

where k_i reflects the model parameters in Equation 5.1.

$i = 0$ to n.

The series of expansion equations relate Log INDEX to the variation of each factor with the urban/rural nature of counties.

The substitution of the Expansion Equations (Equation 6.1) into the initial model, yields the terminal regression model.

$$\text{LogINDEX} = k_{00} + k_{01} \text{TYPE} + k_{10} \text{ FACTOR1} + k_{11} \text{FACTOR1*TYPE} \text{ .... } k_{n0} \text{ FACTORi} + k_{n1} \text{FACTORn*TYPE}$$

(Equation 6.2)

This terminal model will be used to explore urban-rural variation of the general model of poverty. The terms of the equation that are cross products with the dummy variable TYPE represent the variation in the parameter for urban counties. The terms that are not cross products with type express the parameters for rural counties. The final model selected for use in the analysis of urban-rural variation in poverty was selected by use of a "stepwise" regression option. A stepwise regression allows for selection of the "best" regression model. A sequence of models is generated for each variable level which maximizes the R^2; for a one variable, two variable, up to an n variable model. This allows selection of the model that maximizes the R^2 while keeping all terms in the model significant. Thus, stepwise regression allows re-estimation of the model using only significant parameters, and eliminates those that are not significant. This "best" model, one which includes the maximum number of variables while keeping all included variables significant, is the one used for subsequent analysis. Use of stepwise regression may be criticized as eliminating variables without conceptual justification. It is argued here that use of stepwise regression does little to change the value of the R^2 or model parameters, and does serve to present a clear picture and a model uncluttered by insignificant terms. These contentions are supported by presentation of the results of both the full and stepwise models.

Regional Variation. To investigate the regional variation of poverty, eight dummy variables were created to identify which of the nine geographic divisions a county is located in. When all dummy variables are set to zero this indicates a county is located in the South Atlantic geographic division. A value of 1 for dummy variable D1 indicates a county is located in the Pacific geographic division. A value of 1 for dummy variable D2 indicates inclusion in the Mountain geographic division. Dummy variable D3 indicates inclusion/non-inclusion in the West South Central geographic division, D4 in the East South Central geographic division, D5 in the West North Central geographic division, D6 in the East North Central geographic division, D7 in the Middle Atlantic geographic division, and D8 in the New England geographic division (Figure 20).

As in the previous investigation of urban/rural variation in poverty, regional variation is examined via utilization of the Expansion Method (Casetti 1972).

The initial relationship between income poverty (INDEX) and the concomitants of poverty and characteristics of the poverty population was again represented by the model presented in Equation 5.1.

The differential effect of the 12 factors on INDEX, by census geographic division, was investigated using a series of expansion equations. Parameters in the initial model were expanded by each of the eight dummy variables D1 through D8 as follows:

$$k_i = k_{i0} + k_{i1}D1 + k_{i2}D2 + k_{i3}D3 + k_{i4}D4 + k_{i5}D5 +$$
$$k_{i6}D6 + k_{i7}D7 + k_{i8}D8$$

(Equation 6.3)

where k_i reflects the model parameters in Equation 5.1.

$i = 0$ to n.

Each expansion equation relates Log INDEX to the variation of a factor with geographic division.

The substitution of the expansion equations into the initial model yields a terminal regression model that includes terms for the intercept and each of twelve factors
for the nine geographic divisions identified. Thus, the terminal model contains 13 * 9, or 117 terms.

$$LogINDEX = k_{00} + k_{01}D1 + k_{02}D2 .. + k_{08}D8 + k_{10}$$
$$FACTOR1 + k_{11}FACTOR1*D1 +....k_{18}FACTOR1*D8 +$$
$$........+ k_{n,0} FACTORn + k_{n,1}FACTORn*D1 + ...$$
$$k_{n,8}FACTORn*D8 \qquad \text{(Equation 6.4)}$$

The final model selected for use in the analysis of regional variation in poverty was again selected by use of a "stepwise" regression option. The stepwise regression of the regional model generates 116 models, which maximize the R^2, for each variable level from 1 to 116. Again the model selected for interpretation is the "best" model which includes the maximum number of variables while keeping all included variables significant.

In order to illustrate the spatial variation by region for each factor a series of maps were produced (Figures 21 - 26). The parameter values for each of the nine geographic divisions gives a profile of each region.

The numerical data used to address this research question were the same as utilized to develop the general national model of poverty. Thus, data were obtained via mainframe computer access of the magnetic tape version of the Bureau of the Census County and City Data Book (U.S. Department of Commerce 1983). Data were obtained for all 3109 county or county equivalents contained within the forty eight conterminous U.S. states in 1980.

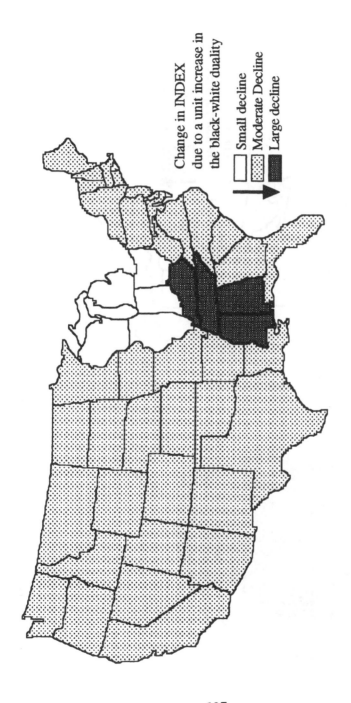

Figure 21. The Effect of a Unit Increase in the "Black-White Duality" on the Income Index, by Geographic Division.

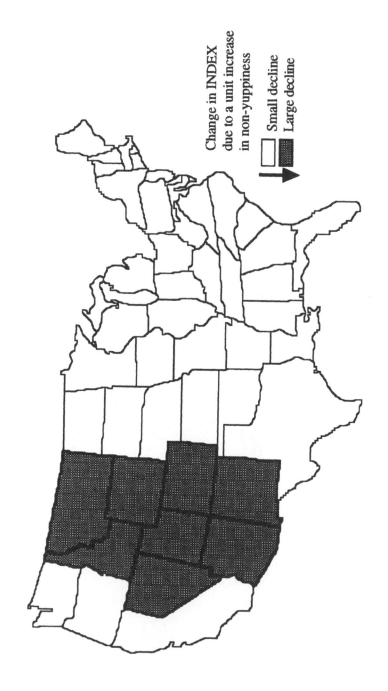

Change in INDEX
due to a unit increase
in non-yuppiness

☐ Small decline
■ Large decline

Figure 22. The Effect of a Unit Increase in "Non-Yuppieness" on the Income Index, by Geographic Division.

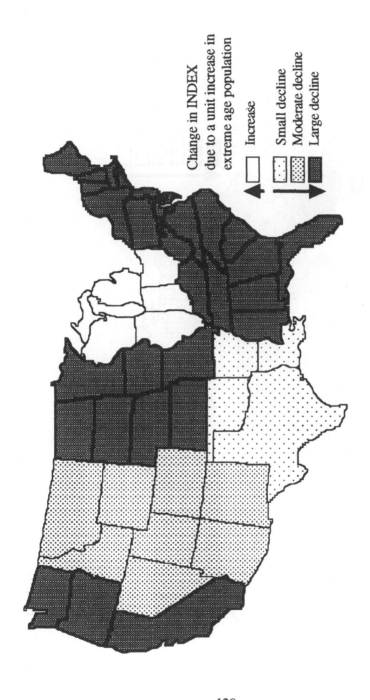

Change in INDEX
due to a unit increase in
extreme age population

→ Increase

☐ Small decline
▨ Moderate decline
■ Large decline

Figure 23. The Effect of a Unit Increase in "Extreme Age Population" on the Income Index, by Geographic Division.

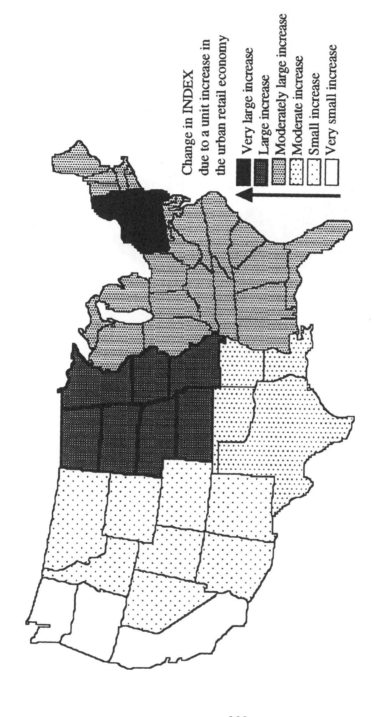

Change in INDEX
due to a unit increase in
the urban retail economy

■ Very large increase
▨ Large increase
▧ Moderately large increase
⬚ Moderate increase
⬚ Small increase
☐ Very small increase

Figure 24. The Effect of a Unit Increase in the "Urban Retail Economy" on the Income Index, by Geographic Division.

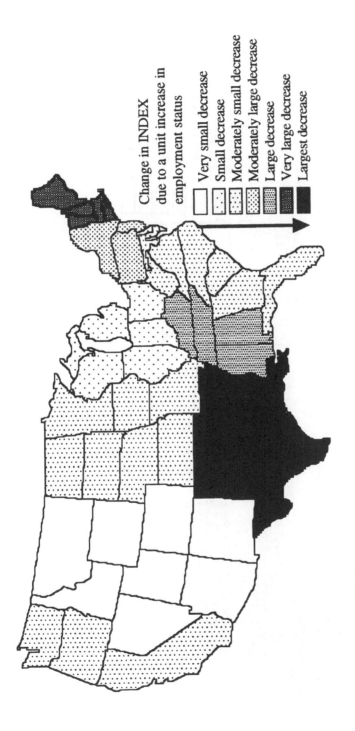

Figure 25. The Effect of a Unit Increase in "Employment Status" on the Income Index, by Geographic Division.

Change in INDEX
due to a unit increase in
employment status

Very small decrease
Small decrease
Moderately small decrease
Moderately large decrease
Large decrease
Very large decrease
Largest decrease

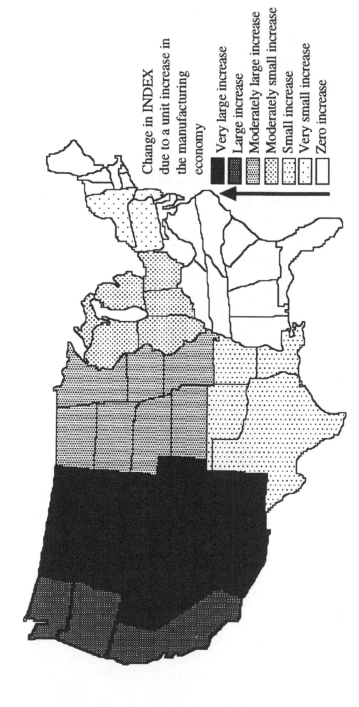

Figure 26. The Effect of a Unit Increase in the "Manufacturing Economy" on the Income Index, by Geographic Division.

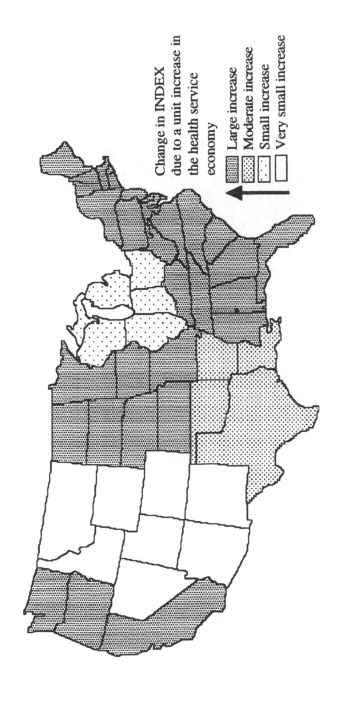

Change in INDEX
due to a unit increase in
the health service
economy

Large increase
Moderate increase
Small increase
Very small increase

Figure 27. The Effect of a Unit Increase in the "Health Service Economy" on the Income Index,
by Geographic Division.

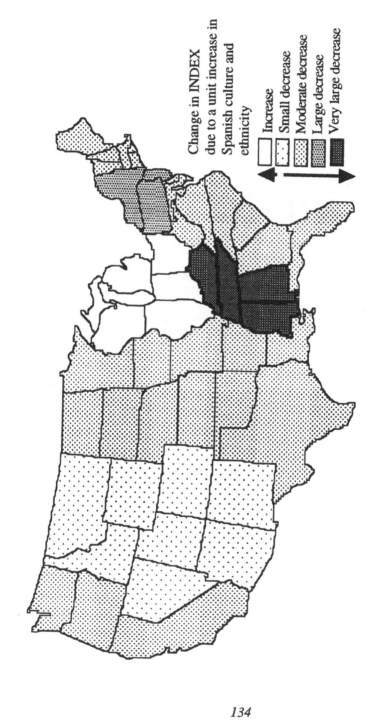

Figure 28. The Effect of a Unit Increase in "Spanish Culture and Ethnicity" on the Income Index, by Geographic Division.

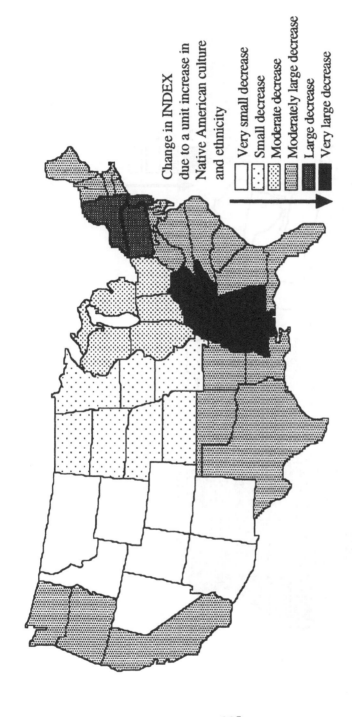

Change in INDEX
due to a unit increase in
Native American culture
and ethnicity

Very small decrease
Small decrease
Moderate decrease
Moderately large decrease
Large decrease
Very large decrease

Figure 29. The Effect of a Unit Increase in "Native American Culture and Ethnicity"
on the Income Index, by Geographic Division.

135

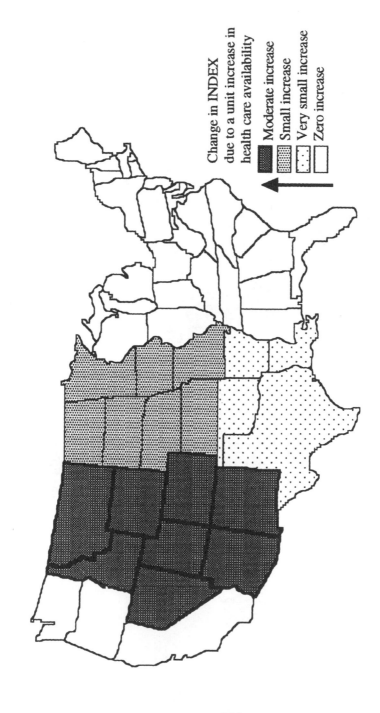

Figure 30. The Effect of a Unit Increase in "Health Care Availability" on the Income Index, by Geographic Division.

Change in INDEX
due to a unit increase in
health care availability

Moderate increase
Small increase
Very small increase
Zero increase

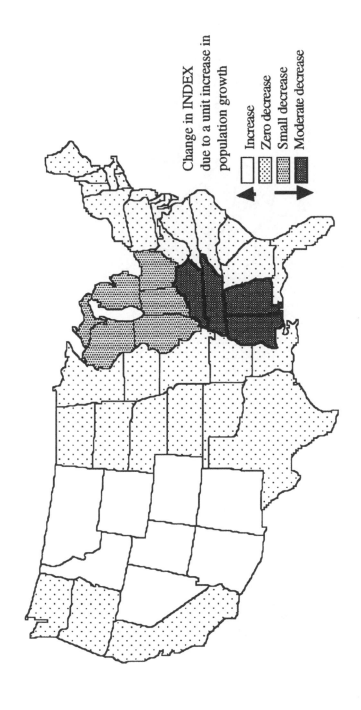

Figure 31. The Effect of a Unit Increase in "Population Growth" on the Income Index, by Geographic Division.

137

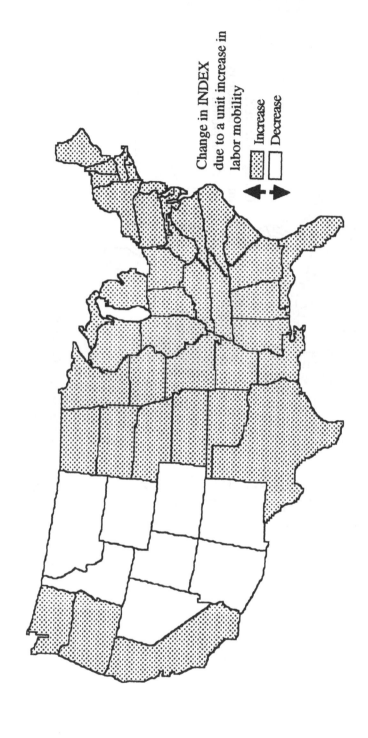

Change in INDEX
due to a unit increase in
labor mobility

� Increase

▢ Decrease

Figure 32. The Effect of a Unit Increase in "Labor Mobility" on the Income Index, by Geographic Division.

Results and Discussion

Factor Maps. A series of twelve maps display spatial variation in the twelve factors generated by the Principal Component Analysis (Figures 7 through 18). Thus, each map illustrates spatial variation at a county level in a dimension of poverty. In the subsequent analysis spatial variation in the impact of factors on poverty are investigated and profiles of the various poverty regions developed. To get a sense of the actual impact of factors, this information must be viewed in conjunction with factor scores at the various locations. While a unit increase in a factor may result in a relatively large change in the income index, the actual importance of the impact must be a function of the magnitude of the level of a factor in each county.

Mapping scores for the black-white duality reveals that high positive scores, which are likely to result in high levels of poverty, are clustered in distinct regions (Figure 7). These high scores can be found in extreme southeastern Utah, northeastern Arizona, and adjacent New Mexico as well as in the Mississippi Delta and the coastal plain from Mississippi to Maryland. There are few areas where black-white duality scores are highly negative; beyond the areas discussed factor scores show little variation.

Mapping non-yuppieness at the county level illustrates that high positive scores tend to be concentrated in the central section of the United States (Figure 8). A broad band of non-yuppieness extends from the eastern Dakotas south to Texas and east to the Mississippi River. Also, a cluster of non-yuppieness is located in Florida and in the Denver area of Colorado. Again, there is no easily identifiable spatial pattern to high negative scores of non-yuppieness.

The high positive scores of extreme age population that are likely contributors to poverty can be seen in one distinct cluster (Figure 9). Extreme age population appears to be concentrated in an area centered in Utah which extends into southern Idaho. Some other smaller areas are evident. One such area is in eastern Arizona-western New Mexico, another spans eastern Wyoming, eastern Montana, and central South Dakota. Other small areas can be seen in west Texas and in a very narrow band that hugs the Mississippi River in the Delta region. High negative scores of extreme age population again show no clear spatial pattern.

High negative scores in terms of the urban retail economy are most likely to be associated with poverty. These areas that lack an urban retail economy have a distinct spatial pattern (Figure 10). Central

Nevada, the four corners region, as well as eastern Montana to western portions of the Dakotas all have highly negative factor scores. In addition, a prominent area where retail economy is lacking can be seen in the central Appalachians. The boundary of this area with the region to the north that has much higher levels of retail activity is very distinct and runs along the northern borders of Kentucky, West Virginia, and Virginia. High positive scores, which indicate a concentration of urban retail activity, can be observed in the west coast states as well as in southern Arizona and Southern New Mexico.

High positive scores on factor 5, which indicate high levels of unemployment, can be observed in a few areas of somewhat limited extent (Figure 11). However, within these areas it seems unemployment is extremely high. Pockets of high unemployment can be seen in northern Idaho/northwestern Montana, northern California/southwest Oregon, northeastern New Mexico, northern Minnesota, northern and peninsula Michigan, northern New York, northeastern Maine, and southeastern Missouri. In addition, a relatively large area with high unemployment is located in the central Appalachians of eastern Kentucky and West Virginia.

The region, which lacks a manufacturing/blue collar economy and so has highly negative factor scores, extends in a broad band from Arizona, Utah, and Idaho in the west to the Dakotas, Nebraska, Kansas, and West Texas in the east (Figure 12). Other areas that lack manufacturing activity can be seen in the central Appalachian region of Kentucky and West Virginia, as well as in the Mississippi Delta region. Areas with relatively high levels of manufacturing activity can be seen in several locations. These include an area from Illinois to Pennsylvania, southern Michigan, northern Illinois and southern Wisconsin, western Maine, western Kentucky, Tennessee, western North and South Carolina, northeastern Mississippi, northern Alabama and Georgia, western Arkansas, and southern Missouri.

There are several areas within the United States that appear to lack a health service economy (Figure 13). These include much of Nevada, extreme southeastern New Mexico and adjacent areas of West Texas, much of Oklahoma, and extreme eastern Kentucky and adjacent areas of West Virginia and Virginia. Areas with a well developed health service economy are scattered across the country but the most prominent are New Mexico, Arizona, and in the Northeast.

People who are of Spanish culture and ethnicity are concentrated within a few distinct areas of the United States (Figure 14).

Southern California, southern Arizona, most of New Mexico, south-central Colorado, the Texas-Mexico border area, and southern Florida all have high concentrations of people who are of Spanish ethnicity. Most of the remainder of the country shows little variation in levels of Spanish culture and ethnicity.

One distinct cluster of Native American population stands out in northeast Arizona, northwest New Mexico, and extreme southeastern Utah. Other small areas are scattered in the northern United States west of the Mississippi River and in eastern Oklahoma (Figure 15). As with Spanish ethnicity, levels of ethnically Native American population show little variation over the rest of the country.

Lack of health care availability seems not to be experienced in any coherent geographic areas, but instead can be found in small scattered pockets (Figure 16). These pockets are most frequently to be seen in the Rocky Mountain region west of the 100o meridian. The biggest area that lacks health care availability within this general region is located along the front range in Colorado. A small eastern cluster of lack of health care availability is to be found in eastern Virginia and extreme northeastern North Carolina. No clear spatial pattern of high health care availability is discernible.

Lack of population growth, shown by highly negative factor scores, shows a distinct spatial pattern (Figure 17). One region encompasses much of the northern United States from the Dakotas, Kansas, and Nebraska to New England. A second area that lacks population growth is located in northeastern New Mexico, and south central Colorado. High population growth appears to be taking place in several regions. One such region spans California, western Nevada, Arizona, western New Mexico, Utah, northwest Colorado, and southern Wyoming. Population growth is also occurring in east Texas and Florida.

Much of the West north of Arizona and New Mexico appears to lack labor mobility, as do Florida and the central Appalachians (Figure 18). It must be remembered that data concerning crossing county lines is not without its problems. Western counties tend to be large and so crossing county boundaries between work and home is less likely than in the East were counties are smaller. High labor mobility is seen in areas that include New England, New York, Michigan, Virginia, and the Mississippi Delta.

Spatial Variations in the Poverty Model. The full regression model that explores possible urban-rural variation contains 25 variables as indicated in Equation 6.2. The stepwise regression conducted indicates that the "best" model in which all parameters remain significant and which maximizes the R^2 includes 22 variables. The parameters of the model which contains all twenty five variables and the results of the "best" model, as identified by the stepwise regression, are presented in Appendix B2.

The full 25 variable regression model indicates that all factors are significant for rural counties. A majority of factors show significant variation from urban to rural counties indicating that substantial spatial variation exists in the model. However, three factors show no significant difference in their effect on urban counties. Thus, the terms which represent the urban adjustment for factors 3, 7, and 11 are insignificant in the model (Appendix B2).

The "best" model, as identified by the stepwise procedure, contains 22 variables. Absent from the model are the urban terms for factors 3, 7, and 11, the same three terms which appeared insignificant in the full model. Use of the stepwise procedure confirms that the effect of extreme age population, health service economy, and growth does not vary between urban and rural counties. Removal of these terms does little to affect the power of the model, as reflected by the minimal reduction in the R^2 from 0.7284 to 0.7281.

Looking at the model results for rural America reveals that four factors exert a relatively powerful influence on LogINDEX. Non-Yuppieness (Factor 2), the black-white duality (Factor 1), employment status (Factor 5), and the urban retail economy (Factor 4) are the most influential factors. The three most powerful factors (non-yuppieness, the black-white duality, and employment status) depress levels of adjusted per capita income. Thus, in the rural United States non-yuppieness, African American population, and unemployment are all associated with income poverty. A positive influence on income is exerted by the existence of any urban retail economy in rural areas.

Two factors, the health service economy (Factor 7) and Spanish culture and ethnicity (Factor 8), are the next most influential on adjusted per capita income in rural counties. The existence of a health service economy positively affects LogINDEX, while a population that is ethnically Spanish seems to depress LogINDEX.

Extreme age population (Factor 3) and Native American ethnicity (Factor 9) appear to have a small but significant negative effect on the dependent variable. In contrast, manufacturing or health care economies, as well as labor mobility, have a small positive effect.

The least influential variable on adjusted per capita income is growth (Factor 11). While the influence of growth in rural areas is small, it is still positive and statistically significant.

The results presented above for rural America appear to largely mirror those presented for the nation as a whole. However, there are some minor differences. The existence of a manufacturing economy has slightly less positive influence on income in rural counties than for the nation as a whole. In contrast, extreme age population has a more negative effect and health care availability a more positive effect in rural counties when compared to U.S. counties as a whole.

The urban parameters presented in Table 7 indicate that the two largest urban-rural differentials are associated with Factors 6 and 9. A manufacturing economy has a greatly increased positive impact, and Native American ethnicity a much more negative impact, on income in urban as opposed to rural counties.

The urban retail economy (Factor 4) and employment status (Factor 5) also show a large urban-rural differentiation. Not surprisingly, variation in the extent of the urban retail economy influences adjusted per capita income less in urban than rural counties. In contrast, unemployment has a more profound impact on income in urban as opposed to rural counties. Spanish ethnicity (Factor 8) and labor mobility (Factor 12) also exhibit a relatively large percentage change when applied to urban counties. It seems that Spanish ethnicity in an urban county, is more strongly associated with low income than in a rural county. Labor mobility is less important to income in urban than rural counties.

The black-white duality (Factor 1) and non-yuppieness (Factor 2), while relatively powerful in both the national and rural models, shows little urban-rural differentiation. However, the negative impact of blackness is slightly accentuated in urban counties when compared to rural America. The effect of non-yuppieness is somewhat less in urban compared to rural counties.

The urban-rural difference shown in health care availability (Factor 10) reveals an unexpected switch. While health care availability is positively associated with LogINDEX in rural counties, the direction of the relationship is reversed in urban counties. It seems that health

Table 7. Recomputed Urban and Rural Parameters from
Appendix B2.

Variables	Parameter Estimates		
	Rural	Urban	Ratio Urban/Rural
Intercept	8.2018	8.2627	1.0074
Black-White Duality	-0.1049	-0.1270	1.2106
Non-Yuppieness	-0.1233	-0.0949	0.7697
Extreme Age	-0.0166	-0.0166	1.0
Urban Retail Economy	0.0950	0.0566	0.5958
Employment Status	-0.0961	-0.1269	1.3205
Manufacturing Economy	0.0163	0.0645	3.9571
Health Service Economy	0.0356	0.0356	1.0
Spanish Ethnicity	-0.0329	-0.0472	1.4347
Native American Ethnicity	-0.0186	-0.0718	3.8602
Health Care	0.0161	-0.0044	0.2733
Growth	0.0079	0.0079	1.0
Labor Mobility	0.0380	0.0197	0.5184

care availability in urban counties is associated with somewhat depressed levels of adjusted per capita income.

The above results for the model that investigates the variations in poverty from urban to rural counties provide a different picture of urban America when compared to that presented for the nation as a whole and for rural counties. In rural America non-yuppieness, levels of African American population, and unemployment are clearly the characteristics mostly closely associated with income poverty. In addition, the existence of an urban retail economy is important to a rural county's economic health. Manufacturing has less of an effect on the economy of a rural county than might be anticipated for the nation as a whole, indicating that the type of manufacturing attracted to rural counties probably has low economic benefit.

While some factors show stability across urban and rural areas, others exhibit distinct differences in their relationship to income poverty. It appears that extreme age population is associated equally with poverty in both urban and rural counties, and the extent of the health service economy and population growth show no urban-rural differentiation. It is in the area of manufacturing that perhaps the most crucial rural-urban differential emerges. Manufacturing in urban counties imparts substantially more economic benefit than it does in rural counties. It may be that the differential is a product of the type of manufacturing likely to be attracted to urban versus rural areas. Much of the manufacturing operating in rural counties requires unskilled labor. The main reason for a rural location is to take advantage of cheap, often female and non-union labor. Examples are provided by the textile and food processing industries. The cutting and sewing of garments and poultry processing are common activities in small communities across much of rural America. This work requires limited skills and little formal education and is low paid. The stronger link between affluence and manufacturing in urban areas supports elements of the shift from manufacturing to services structural explanation of urban poverty. If jobs are within services rather than manufacturing in an urban county it appears to result in an increased likelihood of poverty. The power of this particular structural thesis is greatly reduced for rural counties probably because of the varying nature of manufacturing.

Another interesting urban-rural contrast is in the effects of unemployment on poverty. Despite higher utilization of the welfare system in urban areas, the link between employment and well-being is clearer in urban areas. Contrary to conservative assertions, the choice not to work and dependency on welfare is not likely to be a pleasant experience in urban counties. One explanation of this urban-rural differential might rest on the extent of informal economic activity in rural areas. It has been this researcher's experience that substantial income is generated in rural areas outside the formal workplace. Examples of such informal activities are carpentry, house repair, craft and produce vending, and a variety of services that serve the local market. Some of these activities also aim to draw income from outside the local area from tourists and travelers. Many informal economic activities operate within a local social network distributing the income that is formally earned through the community. This informal activity and flow of money may ameliorate the effects of employment status on income in rural counties. An additional consideration which may weaken the apparent link between employment and income, is the reference week used by the Bureau of the Census. Especially within employment linked to agriculture and recreation and tourism work is very seasonal. The whole of a person's income may be made during a portion of the year. This is especially true within the tourist industry. Thus, a person may report as unemployed during the reference week but have substantial income at other times during the year.

Lack of health care availability is clearly a concomitant of poverty only in rural areas. Health care services are concentrated in urban centers while some rural counties lack basic access. There are many rural counties that have no hospital beds available. This finding may give a somewhat erroneous picture. Just because the urban poor live in a county where there is a concentration of hospital beds and doctors does not mean they receive adequate health care. However, a lack of doctors and hospital beds in rural counties is a good indication of inadequate health care services. A county with no hospital obviously cannot offer emergency care. In addition, it cannot provide routine facilities for such things as childbirth. Lack of such routine services can have serious and lasting health consequences.

It seems that race and ethnicity are more of an issue in urban than in rural America. African American, Spanish ethnicity, and Native American status are all more strongly associated with poverty in urban counties. One reason for this may be because in urban areas these racial and ethnic groups form an easily identifiable minority that may face labor market discrimination. In rural America African Americans, those of Spanish ethnicity, and Native Americans tend to form majorities in the counties where they live, and thus, may not face such intense discrimination. Surprisingly, it is not African American status that shows the clearest urban-rural distinction as might be expected. Rather, it is Native American ethnicity which shows the largest variation in its relationship to poverty when comparing urban and rural counties. An explanation for this link between urban poverty and Native American ethnicity may be found in the nature and location of the Native American population. High percentages of Native American population are seldom found in urban counties but the highest urban levels are regionally very localized. In the larger cities where economic opportunities are most abundant the population that reports itself as Native American is very small. Urban Native Americans seem to be regionally ghettoized much as are their rural counterparts. The urban areas in which this population concentrates are not those which offer the best opportunity for affluence.

It is possible that the generally stronger link between race and ethnicity and poverty in urban areas is to be found in the diversity of urban environments that exist in close proximity. Whereas rural counties within a region may all offer similar environments, urban counties do not. When establishments that provide jobs locate, they may avoid impoverished counties in favor of more affluent counties close by. Poor urban counties may have crumbling infrastructure, substandard buildings, a small local market potential, poor quality of education, health care and other services, as well as a host of social and environmental problems. Neighboring, more affluent, counties may offer a much more attractive location for all kinds of business which provides jobs. The polarity that exists within urban America may thus be compounded along already established racial and ethnic lines.

The full regression model that investigates regional variations in poverty contains 116 variables as indicated in Equation 6.4. The stepwise regression conducted eliminates a substantial number of variables and indicates that the "best" model in which all parameters remain significant and which maximizes the R^2 includes 56 variables. Appendix B3 gives the results of both the full and stepwise models.

The parameters reported by the stepwise model were depicted by a series of maps in order to illustrate regional variation in the influence of each factor on the income index (Figures 21 - 26). It is clear from the maps that while some factors vary little by region, others vary across a majority of the nine divisions and are clearly spatially less stable. All factors, even those that are relatively spatially stable, show some significant spatial variation in their influence on the income index. Thus, not only is there variation between urban and rural counties, but there are also some distinct regional variations. Several regional variations in the concomitants of income poverty and the characteristics of the poverty population stand out. These regional variations may have important implications for the identified poverty cores.

The influence of the black-white racial duality (Factor 1) is relatively stable spatially. However, two regions show significant variation from the majority of the country (Figure 21). In the East South Central region, which includes Alabama, Mississippi, Tennessee, and Kentucky, a high percentage of African American population is more strongly associated with low income than in other regions of the United States. These general findings for the region are clear, despite the inclusion of Kentucky where the previously identified Appalachian poverty core is not associated with high levels of black population. A contrasting impact of the black-white racial duality is revealed in the East North Central geographic division. In Illinois, Indiana, Michigan, Ohio, and Wisconsin the percentage of African American population has less impact on the income index than in the rest of the nation. However, the impact of the black-white duality, although diminished, remains negative. Table 8 clearly shows that in the East South Central census division the black-white racial duality is one of the most important factors that is linked to a high incidence of poverty. In the East North Central division the importance of this racial duality is overshadowed by other factors.

These results, when viewed in conjunction with factor scores, indicate that the black-white duality is important in a distinct but quite limited area. Essentially, this area includes the lower Mississippi Valley, mid sections of the coastal plain from Mississippi to South Carolina, and coastal North Carolina to Delaware. Regional variations in the strength of the impact of this duality indicate that it is particularly crucial to understanding poverty in the Mississippi valley and coastal plain Mississippi and Alabama. These areas are quintessentially Deep South. The history of this region is founded on the overt inequality of the races (Coleman 1991). This inequality was an essential part of the plantation agriculture of the antebellum South but it did not end with the abolition of slavery. The African American population was ill prepared for "freedom." They lacked land, tools, and capital and, as a result, a new dependency developed (Ford 1973). The

Table 8. Recomputed Regional Parameters from Appendix B3.

Variables	Parameter Estimates		
	South Atlantic	Pacific	Mountain
Intercept	8.2122	8.3914	8.2122
Black-White Duality	-0.1151	-0.1151	-0.1151
Non-Yuppieness	-0.0976	-0.0976	-0.1281
Extreme Age	-0.0293	-0.0293	-0.0028
Urban Retail Economy	0.0828	0.0284	0.0487
Employment Status	-0.1119	-0.1119	-0.0624
Manufacturing Economy	0.0	0.1081	0.1175
Health Service Economy	0.0490	0.0490	0.0217
Spanish Ethnicity	-0.0504	-0.0504	-0.0325
Native American Ethnicity	-0.0665	-0.0665	-0.0008
Health Care	0.0	0.0	0.0310
Growth	0.0	0.0	0.0195
Labor Mobility	0.0301	0.0301	-0.0043
	W South Central	E South Central	W North Central
Intercept	8.2122	8.1708	8.1489
Black-White Duality	-0.1151	-0.1656	-0.1151
Non-Yuppieness	-0.0976	-0.0976	-0.0976
Extreme Age	-0.0019	-0.0293	-0.0293
Urban Retail Economy	0.0667	0.0828	0.1105
Employment Status	-0.1896	-0.1547	-0.1119
Manufacturing Economy	0.0280	0.0	0.0631
Health Service Economy	0.0241	0.0490	0.0490
Spanish Ethnicity	-0.0504	-0.1482	-0.0504
Native American Ethnicity	-0.0665	-0.1955	-0.0092
Health Care	0.0156	0.0	0.0181
Growth	0.0	-0.0416	0.0
Labor Mobility	0.0301	0.0301	-0.0301

Table 8. Continued.

	E North Central	Middle Atlantic	New England
Intercept	8.3204	8.2122	8.1291
Black-White Duality	-0.0389	-0.1151	-0.1151
Non-Yuppieness	-0.0976	-0.0976	-0.0976
Extreme Age	0.0216	-0.0293	-0.0293
Urban Retail Economy	0.0828	0.1454	0.0828
Employment Status	-0.0953	-0.1410	-0.1619
Manufacturing Economy	0.0546	0.0274	0.0
Health Service Economy	0.0235	0.0490	0.0490
Spanish Ethnicity	0.0620	-0.1372	-0.0504
Native American Ethnicity	-0.0335	-0.1908	-0.0665
Health Care	0.0	0.0	0.0
Growth	-0.0237	0.0	0.0
Labor Mobility	0.0301	0.0301	0.0301

era of reconstruction established a system of segregation and discrimination that has fundamentally created a southern rural caste system that survives today. Black and white southerners are not socially or economically equal, and racial status constrains choices about such things as marriage partner, occupation, and residence location. Although racial discrimination and segregation are now illegal, the rural caste system continues to operate in a multitude of ways that are often recognized by both black and white. Racism is common in both individuals and within institutions. Culture, values, attitude, sources of self-esteem and pride, sources of social censure and control, family dynamics, gender roles, social networks, and general life experience are different for black than for white southerners (Reul 1974). The black-white racial duality is a complex and deeply entrenched phenomenon in the rural South. Addressing these racial issues is not simply a matter of eliminating overt discrimination and segregation. Perhaps some closing of racial gaps can come through change within both racial groups and not by an expectation that African Americans must conform to white culture and values in order to enjoy economic comfort.

Undoubtedly, capitalism feeds on the South's deeply entrenched racial divisions. However, racism at both the individual and institutional levels, appear to be what sets this region apart from the rest of the country and allows the black-white racial duality to be such a powerful factor in poverty status. A major component of the continued racial duality in the South is the fact that it is deeply entrenched in institutions. For example, education remains unequal despite efforts, both cosmetic and sincere, at equalization. Georgia provides an example of southern racial inequality in education.

Counties in Georgia form county school districts that are substantially dependent on local funding. Counties with primarily black populations have, in general, relatively poorly funded schools. Teachers in these counties are paid less and are less educated and experienced. The result is that students in counties with large African American populations have a lower standard of achievement as measured by standardized tests (Zhou and Shaw 1993). This is not only a result of unequal funding but also of the system of educational control. In addition, this racial difference in standardized test scores may be compounded by cultural bias within the tests. Education as an institution in predominantly black counties is often controlled by the minority white population. The children of those who control the public schools frequently attend private schools. Thus, those who control education have an interest in keeping costs at a minimum but little interest in improving black education (Spring 1985; Hepburn 1992).

The education system is also suffused with more subtle inequalities. An example was provided during an interview with a Clarke county, Georgia teacher in 1992. This educator expressed a different expectation for black than for white students. Teacher expectation can have a profound effect on student performance. This same teacher also saw the task of preparing students differently depending upon their race. The type of jobs black students are likely to do as adults apparently influenced the perception of what constituted an appropriate education. Admittedly, this interview was one opinion; however, it was expressed by an educator in what might be considered as one of Georgia's more progressive counties. It seems likely that others deeply involved with the education of southern African Americans hold similar views.

Such views are not confined to education. In Georgia, in order to get AFDC payments clients must participate in the "Peach" training program. Each client is assessed as to capability and needs. Within the office that administers AFDC in Georgia, a senior official characterized black clients as lacking motivation and educational potential. With such attitudes, its seems unlikely that African American clients are encouraged to fulfill their true potential.

Systematic racism on the part of individuals should not be discounted as a force contributing substantially to the black-white racial duality in the South. It is individuals who often make hiring decisions. As mentioned in a previous chapter, overt prejudice is often supplemented by negative stereotypes of African Americans as potential employees. In addition, hiring of African Americans may meet with resistance from existing employees. Employment segregation is evident in the South even to the most casual observer. Social segregation is also obvious at every level and is an indication that individuals, even if not discriminatory, do feel a racial separation.

Figure 21 shows that there is little spatial variation in the negative effect of non-yuppieness on the income index. Only the Mountain geographic division appears to experience a significantly different impact. In the eight mountain states a lack of what might be considered a "Yuppie" population when viewing county level populations, can be expected to lower the income index more than in the remainder of the United States. In the Mountain region lack of a

Yuppie population is the dominant negative impact on income within the model. However, most areas that lie within the Mountain region show no concentration of non-yuppie population. The small core of non-yuppieness in the Denver region is an exception. There is no indication that the southwestern poverty core, which is located in the Mountain region, has any particularly significant Non-yuppie population. It appears that Non-yuppieness is a dispersed phenomenon and its importance to poverty has no clear regional dynamic. Thus, promotion of a college education and efforts to maintain a population balanced by a significant segment of young adults may be equally effective in all regions.

The effect of extreme aged population on the income index shows significant variation over more regions than either the black-white racial duality or non-yuppieness. Moreover, the effect of extreme aged population is not always negative as the general model of poverty seemed to indicate (Figure 22). In a majority of regions (the Pacific, West North Central, East South Central, South Atlantic, Middle Atlantic, and New England), the existence of high levels of extreme aged population has a highly negative impact on the income index. In contrast, the East North Central geographic division is revealed to be an anomalous region where extreme aged population has a surprising positive impact on income. In addition to this regional variation in the ability of extreme aged population to depress income, the extreme aged population shows a clear spatial pattern as discussed previously. The combined effect of these conditions is to make extreme aged population a particularly significant factor in two poverty cores: the southwestern core and the core centered in South Dakota. In these cores there are large numbers of the very old and young. In the South Dakota core in particular extreme aged population appears to have a relatively powerful capacity to lower the income index.

One likely reason for these high levels of extreme population is out-migration of the adult population. Both these poverty cores are associated with Native American populations and are, in part, on reservation land. Urbanization is extremely limited, as are job opportunities, and so it would be hardly surprising if young adults left the area. However, viewing the profiles of several counties that are part of the South Dakota poverty core offers no confirmation that out-migration is the cause of high rates of extreme age population. In Shannon and Ziebach counties, for example, it is young children who clearly comprise a large portion of the extreme aged population. In addition, male/female ratios show no indication that young males are leaving in large numbers. In contrast to much of the rest of the state, these poor counties gained population in the 1970s. Instead of out-migration, high levels of extreme aged population appear to be a result of families with unusually large numbers of children. This is evidenced by the reported number of persons per household. The state average is 2.74 for South Dakota, while the average household sizes in Shannon and Ziebach counties are 3.85 and 4.84 respectively (U.S. Bureau of the Census, 1983). A similar situation exists in the southwestern poverty area. Thus, it seems it is the financial drain of young children that is responsible for the link between extreme aged population and low income, rather than out-migration.

Although southern Utah is the only part of that state included in a poverty core (the southwestern), the whole state stands out as having extremely high levels of extreme aged population. Thus, this factor is particularly important to understanding poverty in Utah. Again, closer examination of county statistics indicates a high percentage of very young children in the population. It seems probable that Utah's distinctive demographic structure is a product of the social system that accompanies the dominant Mormon religion with its emphasis on children and the family, and its acceptance of polygamy.

It seems clear that in both poverty cores that have large Native American populations, family structure contributes to financial pressures. A family with more than the average number of children to support obviously needs a higher income to avoid being classified as poor. This family structure that prevails in both the southwestern and Dakotan poverty cores is rooted in traditional Native American culture. Traditional culture, including the valuing of large families, is clung to like a life-line by concentrations of Native Americans, especially on the reservations. In some respects Native American poverty cores represent islands apart from mainstream culture and the capitalist economy of the rest of the country (Harrington 1984).

Figure 22 displays the extensive regional variation in the positive impact of the urban retail economy. The urban retail economy is one of the least spatially stable impactors on poverty. In the Pacific region, the existence of an extensive urban retail economy within a county has only a small positive impact on the income index. In contrast the levels of urban retail economy have a highly positive effect on income in the Middle Atlantic geographic division. In fact, in the Middle Atlantic region the Urban Retail Economy is the most powerful factor that elevates income, while in the Pacific region this factor is one of the weaker positive influences.

If both the regional variation in the effect of an urban retail economy and the spatial distribution of the relevant factor scores are considered, lack of such an economy is clearly an important contributor to poverty in two of the identified cores. The differential regional effects indicate that both the South Dakota and central Appalachian regions are particularly negatively affected. It is in the Appalachian region that lack of an urban retail economy is most widespread. The Appalachians, it has been noted, is relatively densely populated but most of this population is rural. This situation is often attributed to the rugged terrain, the immature transportation network, and the social history of the inhabitants (Appalachian Regional Commission 1972). As has been noted earlier, much of the South Dakota poverty core is reservation land, and even a cursory glance at any map of the U.S. urban network reveals sparse development in the region. Figure 19 confirms the South Dakotan poverty core as essentially rural. As is the case in Appalachia, the road network is immature. There is no interstate or four-lane highway serving the area, and even state highways are relatively widely separated. The lack of infrastructure development and the traditional rural lifestyle of Native Americans combine to produce a very limited urban retail economy that has little power to contribute to the economic health of the region. Both the Appalachian and South Dakotan poverty regions appear to suffer from isolation that is both physical and social. Both remain largely rural in an economy where urban centers appear to be the focus of economic activity associated with relative affluence. However, one major difference between these two cores is in their position within the larger economic structure. Whereas the Dakotan core has no clear role within the national economy, the same is not true of central Appalachia. Appalachia is frequently characterized as having a well integrated position within American capitalism (Harrington 1984). Much has been written concerning the exploitation of Appalachia for its raw materials, particularly coal and lumber, and its low cost labor by absentee capitalists (Wells 1977; Pudup 1987).

The spatially variant effect of employment status is shown in Figure 23. Clearly, employment status has a highly variable impact on the income index dependent upon region. When considering the negative impact of unemployment on income, the nation can be divided into seven areas. The extremes are provided by the Mountain and West South Central geographic divisions. In the Mountain region, unemployment has a relatively small negative impact and other factors seem to more significantly lower income. This contrasts to the substantial impact of unemployment observed in the adjacent West South Central division. In Arkansas, Louisiana, Oklahoma, and Texas unemployment is the most serious depressor of the income index. The rest of the country holds a more middle ground but all regions show varying negative impacts of unemployment.

The national distribution of unemployment, combined with regional variations in the effect of the lack of a job, indicates unemployment is a major problem within both the Appalachian and Texas border poverty cores. In Appalachia the power of unemployment to affect income is less profound than in some other regions. However, such high unemployment levels exist that employment status must be viewed as a major link to poverty. As has been discussed, the Appalachian poverty core lacks an urban retail economy that might be expected to provide jobs. The Appalachian economy is narrow rather than diverse; such jobs that do exist are largely within agriculture, forestry, and mining (Appalachian Regional Commission 1966a; Duncan 1986; Kublawi 1986). These areas of employment have historically suffered a series of boom and bust periods and, in general, have been in long term decline (Guinness and Bradshaw 1985). Coal mining in particular has seen a dramatic decline in employment due to both automation and declining demand (Belcher 1962; Millstone 1972).

Many people characterize central Appalachia as a region exploited for its resources and then cast away when no longer useful (Caudill 1972; Gaventa 1980; Seltzer 1986; Szakos 1986). The mining and timber companies that grew rich from the land and labor did little to develop urban centers or any diversity within the economy that would provide jobs. Indeed, mining and timber companies obviously had a vested interest in monopolizing employment opportunities (Ford 1973; Wells 1977; Armstrong-Cummings 1986). The result of the exploitation for the land is a decimated environment (Brooks 1972; Good 1972); for the people it is chronic unemployment and grinding poverty (McNutt 1986; Seltzer 1986). Unemployment is one important facet of a complex situation in place within Appalachia that may be said to be largely a result of its place within American and global capitalism (Millstone 1972; Armstrong-Cummings 1986; McNutt 1986; Szakos 1986).

Unemployment has its most powerful regional effect in the West South Central division where the Texan border poverty core is located. High levels of unemployment, particularly in the southern extent of this core, indicate that it is a very serious problem. This border region is very isolated and is poorly served by transportation routes. The extreme northern portion of the border core may benefit from Interstate 10. Further south are extensive areas that are less well served. Perhaps more important than the isolation is the fact that for each city of any size on the U.S. side of the border a matching city exists in Mexico. It is almost impossible to control illegal border crossings and so any jobs that are available may be competed for by Mexican as well as American nationals. The southern extent of the border core where unemployment is most severe is the most densely populated portion of this core. However, there is also a dense population on the Mexican side of the border. In addition, major Mexican highways feed into this region. An additional factor that may contribute substantially to unemployment in the Texan border poverty core may be the great advantages that employers enjoy by locating on the Mexican side of the border. In a ribbon from Nuevo Laredo to Matamores in Mexico is a concentration of labor intensive economic activity that exploits low wage Mexican labor and which leaves an employment vacuum on the U.S. side of the border.

The manufacturing or blue collar economy might be discounted as an important impactor of income if analysis were limited to the nation as a whole. However, studies of the separate geographic divisions clearly showed a widely varying impact spatially that is masked within national analysis (Figure 23). Just as urban versus rural America showed a crucial difference in the impact of manufacturing, so does the impact vary regionally. In some geographic divisions, manufacturing has no or minimal effect on the income index, while in others the positive impact is very strong. In the two regions (the East South Central and the South Atlantic) that span the southeastern United States, the extent of the manufacturing economy at a county level has no significant impact on the income index. The picture in the western United States is very different. In particular, in the Mountain and Pacific geographic divisions it is the manufacturing economy that exerts by far the most profound positive influence on income.

This variation has great significance for the various poverty regions. In the east and south of the United States a manufacturing/blue collar economy does little to raise income, a situation that has important implications for the Appalachian, Delta, and Texas border poverty cores. In the West, however, an increase in manufacturing activity seems to be associated with rising income. An explanation of this regional differential in terms of the effect of manufacturing on income may lie in the varying history and nature of manufacturing.

Within old industrial areas of the northeast, manufacturing activity the basis of past wealth is clearly no longer associated with economic well-being. In this broad region, traumatic changes have taken place in recent years. Manufacturing plants have closed and jobs have been lost. Industry that remains faces stiff competition, especially from manufacturing abroad. In addition, the northeast often comes to this competition with old plant and equipment and without the monopoly on technical innovation it once enjoyed (Guinness and Bradshaw 1985). Communities in the old manufacturing belt which rely on manufacturing and blue collar activities are often communities in economic crisis.

Manufacturing has been gradually shifting from the traditional northern manufacturing belt to the South and to the West. It might be expected that this movement would bring substantial benefit to these regions. However, it is important to look at the nature of manufacturing that has made these regional shifts. The South offers non-unionized, low wage labor as one of its primary attractions (Appalachian Regional Commission 1966b). Thus, labor-intensive industries which seek to maximize profits and increase competitiveness by lowering wage costs are likely to locate in the South. Southern state and local governments often offer substantial incentives to industry. These include tax concessions, financial aid, assistance with credit, subsidized land, site assembly services, tailored infrastructure, waiving of regulations, and the development of education systems to suit the needs of companies (Shaw 1963; Newman 1972; Netzer 1974; Friedland 1983; Elkin 1987). These incentives involve substantial costs to the communities and states that offer them such that they may outweigh the benefits, if they exist, that incoming industries have to offer. Moreover, once subsidies expire, companies may leave for more lucrative locations (Newman 1972). Examples of manufacturing that has largely relocated to the South include the textile, shoe, and apparel industries. Since the objective of the move is to minimize labor costs, such manufacturing is not likely to bring affluence to workers or the region (Guinness and Bradshaw 1985). The type of manufacturing found in the Appalachian poverty core is clearly of the type that is likely to bring little or no economic benefit. This manufacturing is heavily concentrated in activity related to the timber industry, textiles, agriculture, and mining (Saunders 1969).

The nature of manufacturing that has relocated to, or established itself, in western states is different from that found in the South. The existence of manufacturing regions such as "Silicon Valley" in the West point to the development of high-tech industries in the region. While not all labor is highly paid in these industries, cheap labor is not the major draw. Much of the required labor force must be skilled and well educated. In addition, these industries are often relatively capital intensive, reducing the importance of labor as a factor of production. The value added by manufacturing in western states is lower than that in northeastern and southern states, but manufacturing appears to be qualitatively different in the various regions (Guinness and Bradshaw 1985). This variation is reflected in the economic effects of manufacturing. Given the positive effect of manufacturing in the West, it is tempting to focus on possible benefits for the Southwestern poverty core. However, high-tech industry seems unlikely to get its needs met in this area. Deficiencies include lack of infrastructure and urban centers, and the population is poorly educated. Many changes would have to occur on the Native American reservations before they could become centers for high-tech industry and high paying jobs.

Existence of a health service economy at the county level has a positive influence on the income index, but this influence varies somewhat by region (Figure 24). The influence is relatively weak when compared to other factors, but its strongest benefit is felt in an extensive six region area. Although the existence of a health service economy does affect income positively in the Mountain, West South Central, and East North Central regions, its influence is weaker than in the rest of the nation.

In addition to the relatively limited spatial variation by geographic division, the effect of the health service economy on income is weak. Moreover, the areas which lack a health service economy are not extensive and do not coincide with poverty regions, with the possible exception of the Appalachian poverty core. Despite these two facts, the lack of a health service economy cannot be totally discounted, especially in poverty stricken areas of Appalachia. The Appalachians are part of a conglomeration of geographic divisions where the positive effect of health service economic activity is at its strongest. Low levels of health service related economic activity within the Appalachian poverty core appear to be part of the general lack of economic diversity discussed earlier.

Spatial variation in the influence of Spanish culture and ethnicity, like that of extreme aged population discussed earlier, is dichotomous (Figure 24). Again, the anomalous region in which influence is positive is the East North Central division. In the rest of the nation, the influence of Spanish ethnicity on the income index is negative. In the East South Central and Middle Atlantic regions, the depressant effect of Spanish ethnicity on income is particularly powerful. It is interesting to note that it is not in regions where the population which is ethnically Spanish is concentrated that such ethnicity has its strongest effect on the income index. People of Spanish ethnicity face the greatest economic disadvantage in Alabama, Kentucky, Mississippi, and Tennessee, not in states where the Spanish population is heavily concentrated. This phenomenon seems to be part of a more general influence of race and ethnicity within the East South Central geographic division. These southern states have a history of racial division and discrimination that may have embraced the ethnically Spanish population. This population is a very small minority group in this region, unlike in states where the Spanish population is concentrated and where it often forms a majority. Where ethnically Spanish population is substantial, discrimination on that ethnic basis is less likely. Moreover, language difficulty with English is less of a handicap, and proficiency in Spanish is essential. It should be kept in mind that the importance of Spanish ethnicity in the Delta and Appalachian poverty cores where its strength as a factor is strongest is limited by the small size of relevant population there.

Although the effect of Spanish ethnicity is weaker in the Mountain and West South Central geographic divisions, it is still negative, and the size of this population is substantial in certain areas. Thus, in the Texas border poverty core Spanish ethnicity is perhaps one of the crucial links to poverty. The ethnically Spanish population and border poverty core show a clear spatial conjunction. The southwestern poverty core also extends over areas where the Spanish population is concentrated. No other poverty factor is so clearly associated with economic deprivation in the border core as is Spanish ethnicity. The borderlands form one truly distinct cultural region within the United States (Guinness and Bradshaw 1985). The various factors appear to combine within this cultural region and interact to produce relatively low income.

The influence of Native American culture and ethnicity is unilaterally negative in terms of the income index. However, the power of this factor to depress income varies greatly by region (Figure 25). The greatest influence of Native American ethnicity is seen in the East South Central geographic division. The weakest effect appears to be in the Mountain and West North Central regions. The weak unit effect of Native American ethnicity in these two regions may appear somewhat surprising. It is in these regions that the poverty clusters that appear associated with high percentages of Native American population are located. However, it should be remembered that even a small decline of income with each unit of a factor can be significant when magnified by high factor scores.

Like the population of Spanish ethnicity discussed above, the Native American population is also extremely spatially concentrated. By far the most coherent and largest core is located in the four-corners area of the southwest. Other smaller concentrations can be observed in Oklahoma and scattered throughout the Dakotas and Montana. This clustered location of Native Americans is a product of a recent history of exploitation and persecution. Europeans ejected Native Americans from desirable areas, tricked them into giving up their lands for trinkets or paltry sums of money, broke numerous treaties, and decimated their population through war and disease. The Native American peoples were treated as subhuman, their children taken away to be "civilized", and a systematic and sustained effort made to destroy their cultures (Harrington 1984). The result has been that areas where Native Americans are concentrated are undesirable, environmentally hostile, and isolated. To some extent, Native Americans have become forgotten people. Some attempts have been made to integrate native populations into the mainstream but many of these attempts have been minimal. In addition, many Native Americans prefer to cling to traditional lifestyles on the limited lands that remain to them. Integration into mainstream America is seen as having the too high price of cultural suicide (Guinness and Bradshaw 1985).

Native Americans are the most distinctly separate ethnic group in the United States. This social separation is reflected in economic separation. Guinness and Bradshaw (1985) identify the root of Native American poverty to lie in the poor quality of reservation land. However, some of these seemingly barren lands do contain appreciable natural resources. These resources include natural gas, coal, and Uranium (Guinness and Bradshaw 1985). It seems that thus far, Native

Americans have not allowed exploitation of these resources or chosen to exploit them themselves.

It is interesting to note that the association between Native American ethnicity and low income is at its weakest in regions where this population is concentrated. The relatively weak effect of Native American ethnicity in areas where this population is concentrated may be due to relatively low levels of discrimination. It is obviously more difficult to discriminate where an ethnically distinct population forms a majority and such ethnicity is a norm. Again, as was the case with Spanish ethnicity, it is within the East South Central geographic division where Native American ethnicity has the greatest power to depress the income index. This powerful effect is once more circumscribed by the small size of the Native American population in this region. Despite the relatively weak negative effect of Native American ethnicity in areas where this population is concentrated, the link between Native American ethnicity and culture and poverty is clear in both the southwestern and Dakotan poverty cores.

For a majority of the nine census Geographic Divisions, health care availability has no significant impact on the income index (Figure 25). However, in three regions, there is a small positive impact. It is in the Mountain Geographic Division that the maximum impact of health care availability on the income index is felt.

The southwestern poverty core is located in the Mountain Geographic Division and there is a moderate lack of health care availability. Thus, health care provision appears to be an issue within this core. In addition, it should be remembered that the counties within the southwestern poverty core are large and even if health care is available, it may be located a long distance from some county residents. Clinics often provide basic health care on the reservations which form substantial portions of the southwestern poverty core. Either patients or health care providers may face an arduous journey in a poverty core where communities are frequently very isolated. People in need of health care may not seek it, or may do so only after a health crisis is in progress.

A discussion with a health care provider in the Navajo reservation which is a part of the southwestern poverty core confirmed the difficulties of health care provision mentioned above. In addition, there are sometimes language barriers to be overcome and cultural resistance to modern medical practice and intrusion from the outside. The discussion also revealed that the Native American population there have some special health problems. Perhaps the most major health problem faced within the southwestern poverty core is that of alcohol abuse. This abuse is widespread and has a multitude of social and economic ramifications. Buying alcohol uses money that might be better used elsewhere while at the same time the alcohol consumed reduces an individual's potential ability to generate income. Alcohol related violence is relatively common both within and outside the family. Alcohol consumption brings with it the common range of associated health problems from accidents to liver damage. Perhaps the most disturbing effect is seen in children of mothers who drink during pregnancy. The health worker that provided this view of health problems indicated that Fetal Alcohol Syndrome is widespread. The diminished mental capacity that is part of this syndrome affects an individual for life and, thus, these children will become adults who are ill-equipped to function within their own or the larger society.

Health care availability does not seem to be a crucial issue outside the Mountain division. However, health care is still a concern in other cores of poverty and should not be summarily discounted even when cores are located in regions where there is no direct effect on the income index. An example is provided by the Appalachian core which is located in the East South Central Geographic Division where lack of health care is indicated not to directly affect income. Great strides have been made in health care provision but the situation is far from perfect (Appalachian Regional Commission 1985). Central Appalachia has a long history of lack of health care. Many counties had no hospital in the late 1950s (Hamilton 1962); today 22 of the 55 most intensely impoverished counties still have no hospital, despite the assertion of progress. While the area in Appalachia that lacks health facilities is relatively small, its size does not make the effects less meaningful for residents. The population in central Appalachia often suffer chronic ill health (Ford 1964; Bethell et al. 1972; Appalachian Region Commission 1972; Guinness and Bradshaw 1985). Despite a decline in line with national trends, infant mortality remains somewhat high (Ford 1974; Osborne et al. 1986). Lack of health care has developed as part of the capitalistic development of the region. The coal and lumber companies for the most part showed little interest in the health of workers severely affected by employment conditions (Scalf 1972). It was organized labor that was the single most progressive force in providing health care (Wells 1986), despite many problems of corruption with the unions (Bethell 1972; Gaventa 1980).

Growth is another factor which for most of the nation has no significant impact on the income index. When growth does impact income its effect is not clear. In the Mountain region growth appears to have a beneficial impact on income as was anticipated. However, in the East North Central and East South Central regions growth results in declining levels of income (Figure 26). Thus, for some regions, growth is detrimental to income, for one it is beneficial, and for the remainder it has no systematic effect.

This result indicates that stimulation of growth might benefit the southwestern poverty core but is likely to worsen conditions in Appalachia. It is a commonly expressed opinion that the population of Appalachia is too dense for either the environment or the economy to support. Indeed some have advocated out-migration as a partial solution to this region's problems (Gamble 1966). Certainly, it appears reasonable that population growth in the Appalachian poverty core which lacks the economic diversity and economic structure to provide for the existing population would only add a greater burden to an already distressed region. Fortunately factor scores indicate that no substantial population growth is occurring and, thus, population growth is not an issue within the Appalachian poverty core.

The southwestern poverty core is experiencing population growth in a region where such growth has been identified as exerting a positive influence on the income index. Despite this result for the Mountain geographic division as a whole, it is difficult to envisage how population growth would benefit the southwestern poverty core. The positive impact of growth for the Mountain geographic division is most likely to be more a result of general sunbelt growth in urban nodes and urbanized regions such as the Front Ranges. In these areas, population growth is the product of economic expansion and, thus, has underlying economic support. In the southwestern poverty core, birthrates are high and the average family size is large, both indications that population growth is largely a result of natural increase due to large families. Population growth due to economic in-migration must have very different effects than population growth within an economically deprived existing population.

The final factor, labor mobility, is relatively spatially stable. However, while in much of the nation mobility is linked to higher income, the Mountain region is an exception (Figure 26). In the Mountain region, it appears that working in a county different from that of residence has a somewhat negative impact on the income index.

This varying regional effect of labor mobility on well-being provides an interesting contrast. Travel to work involves costs but in most areas of the country mobility appears to allow workers to command higher income and benefits outweigh the costs of travel. An exception to this is in the Mountain region. In the mountain region, counties tend to be large and the road network relatively limited. In addition, conditions of weather and terrain are often arduous. Thus, movement across a county boundary to work probably involves a longer more difficult journey than in other regions. The costs of such a journey are likely to be high in terms of the type of vehicle needed, operating costs, and also in terms of time.

Outside the Mountain region, in areas where the road network is relatively less developed and the people isolated, lack of mobility is likely to be a crucial problem. It is just such areas where local employment opportunities are probably very limited. Not only does an immature infrastructure limit the daily movement of people in such an area, but it may discourage some employers from locating there. Travel to work is almost inevitable in order for people to command a decent income. While low mobility could be symptomatic of the existence of ample job opportunities locally, this research indicates that this is not the case. Lack of mobility is associated with low income, and increasing mobility must be a focus of policy designed to alleviate poverty. These findings suggest that development of the transport system, such as was undertaken in the Appalachian region, is a legitimate way to address poverty in most poverty cores. The exception is the southwestern poverty area. Development of the transportation network here is likely to do little to help the poor.

When the model parameters are viewed on a region by region basis they provide a profile of each of the nine Geographic Divisions identified. These profiles are provided in Table 8. Regions vary in several ways from each other. Firstly, the magnitude of the impact of factors on the income index varies spatially. Secondly, the relative importance of a factor with respect to other factors varies. Thirdly, the overall number of factors that impact income varies by region. Fourthly, the balance of negative versus positive impactors of income varies between Geographic Divisions. Lastly, the type of factor which impacts the income index may be distinct within a region.

In the South Atlantic region, the magnitude of the impact of the factors on the income index is moderate. However, although several factors suppress income substantially and some have no impact, there are few positive influences on income. The black-white duality and employment status are both powerful depressors of the income index, and non-yuppieness as well as Spanish or Native American ethnicity also serve to lower income. Thus, it seems to be individual characteristics of the population that are linked to poverty. In contrast, it is the extent of the urban retail economy which exerts the strongest beneficial influence on income.

In the Pacific region the factors that negatively impact the income index are very similar to those in the South Atlantic region. It is in the positive impactors of income that a major contrast exists between these east and west coast areas. In the South Atlantic region the extent of the manufacturing or blue collar economy has no significant impact on the income index, while in the Pacific region it is crucial to high income. In California, Washington, and Oregon the manufacturing economy is the dominant factor associated with relative affluence.

The Mountain region is far from typical of the United States as a whole. All the twelve factors identified have some impact, either positive or negative, on the income index in this region. Moreover, in all factors except the black-white racial duality, the impact differs from that in the rest of the nation. This contrast with the rest of the country is extreme in the case of growth and also labor mobility. The impact of these two factors is in the reverse direction to that observed for the rest of the country. In the Mountain region, the black-white racial duality is strongly associated with low income just as it is in much of the rest of America. However, in the mountain region these racial factors are overshadowed by the negative impact of non-yuppieness. Just as in the Pacific region it is the extent of the manufacturing economy in the Mountain region which is the most powerful counterbalance to factors which depress the income index. While extensive urban retail economies and health service economies within counties do serve to elevate incomes somewhat, the effect is weaker than in many other regions.

The West South Central Geographic Division is one region within which the black-white racial duality is not the most powerful factor in lowering the income index. Instead it is unemployment that is most strongly associated with low income. It is noticeable that all the factors that significantly elevate income are weak when compared to other regions. It is the urban retail economy that most influences income upwards but the effect is less powerful that in much of the rest of the country.

The East South Central region is notable largely because of the very profound impact of race and ethnicity on the income index. In this region the black-white racial composition as well as both Spanish and Native American ethnicity have stronger links to low income than in any other region. It is Native American ethnicity that has the most profound negative association with income. The magnitude of this effect is the greatest for any factor in any region. Along with the powerful income lowering effects of unemployment and Non-yuppieness, these racial and ethnic depressors of income form a formidable group. Ranged against these factors that can be expected to lower income, are levels of the urban retail and health service economies as well as labor mobility. However, the positive impact of these factors does not stand out as being especially strong in the East South Central region. In contrast to the Mountain and Pacific regions, manufacturing is not significantly linked with any change in the income index.

In the West North Central Geographic Division black-white duality and employment status are the two most powerful factors in their ability to lower the income index. It is worthy of note that the impact of both Native American and Spanish ethnicity in the West North Central region, while still negative, is comparatively weak when compared to other Geographic Divisions. It is the extent of the urban retail economy that clearly has the most potential to raise the income index in this region.

Within the East North Central Geographic Division the negative effect of the black-white racial duality is much weaker than in other regions. Like the Mountain region, the East North Central region contrasts with the rest of the nation in several ways. The factors that depress income are fewer and their effect weaker in this region. In fact, in the case of both extreme age population and Spanish ethnicity, the relationship with income is the reverse of that seen elsewhere in the United States. Both these factors have a positive impact on the income index. The income index is lowered most effectively by non-Yuppieness and unemployment. In common with much of the country, however, this region is most positively impacted by the extent of the urban retail economy in place at a county level.

The Middle Atlantic region is influenced by a powerful array of factors that substantially depress the income index. The three factors which constitute the racial and ethnic makeup of the population, the black-white duality, Spanish ethnicity, and Native American ethnicity, are all strong negative influences on income. As in the East South Central region, these factors combine with employment status and non-yuppieness into a formidable block of potential income damaging factors. Unlike the East South Central region, however, the counterbalancing factor of the urban retail economy is more powerful than in much of the nation. Still, it seems the Middle Atlantic region is dominated by negative factors whose potential to depress income is far greater than factors that might be expected to raise income.

The final region, New England, is most notable for the fact that three factors have no significant impact on the income index. As is the case in the South Atlantic region, the manufacturing economy, health care availability, and growth have no impact. The profile that is presented of New England is very similar to that of the South Atlantic region. A minor difference exists in the power of unemployment to depress income. In New England unemployment has a somewhat more negative effect on the income index than in the South Atlantic region.

Some general comments may be made that appear to apply to most regions. The profiles presented for the nine census Geographic Divisions appear to be dominated by negative factors. In all but the East North Central Division there are more negative than positive factors. In every region the most powerful negative factors have a greater per unit impact on the income index than do the strongest positive factors.

TEMPORAL VARIATION IN THE LOCATION OF POVERTY CORES AND THE NATURE OF POVERTY, 1980 TO 1990

Data and Methodology

Changes in United States poverty that occurred from 1980 to 1990 were investigated in three stages:

1. The location and extent of poverty cores in 1990 was explored. The general spatial changes that have occurred in the distribution of United States poverty over the decade of the 1980s was investigated by mapping the adjusted per capita income for 1990 and comparing to the map for 1980 generated previously. These maps show the position and extent of poverty cores in the two years. In addition,

the poorest 5 percent of counties are displayed on Figures 5 and 27 These two maps clearly show the spatial distribution of America's poorest counties in 1980 and 1990.

The statistical analysis aimed at investigation of changes in the impact of important concomitants of income poverty and general characteristics of the poverty population from 1980 to 1990 was accomplished using steps similar to those used to explore regional variations in poverty.

2. Principal components analysis was used once again to group the 16 independent variables into categories. All variables used to model 1980 poverty, which were also available for 1990 were used in the investigation of temporal variation in the nature of poverty from 1980 to 1990. Since some variables available at the county level for 1980 are not yet available for 1990, the temporal model of poverty is necessarily more limited than the spatial models presented earlier. However, sufficient comparable variables are available to adequately represent several major dimensions of poverty and characteristics of the poverty population. These were HSCHOOL, COLLEGE, PLUMB, OVERCR, COMMUT, URBAN, WHITE, BLACK, INDIAN, SPANISH, TEENS, OLD, PERFEM, LABFEM, and UNEMPL (Table 1).

The data were obtained via mainframe computer access of two Census Bureau data sources. Again, 1980 data were obtained from the magnetic tape version of the Bureau of the Census County and City Data Book (U.S. Department of Commerce 1983). Data were obtained for all 3109 county or county equivalents contained within the forty eight conterminous U.S. states in 1980. 1990 county level data were obtained from Summary Tape File 3 (STF3) which is a preliminary compilation of statistical information obtained during the 1990 census. Data were obtained for all 3111 county or county equivalents contained within the forty eight conterminous U.S. states in 1990.

The data for both 1980 and 1990 were pooled, resulting in a total of 6218 observations which were used to generate the factors. All the independent variables pooled were expressed as percentages and so comparability from 1980 to 1990 was not a problem. A Varimax rotation was again applied and an eigenvalue of one used as the retention criterion for factors generated.

3. Changes in the impact of important concomitants of income poverty and general characteristics of the poverty from 1980 to 1990 were modeled using the expansion method. The independent

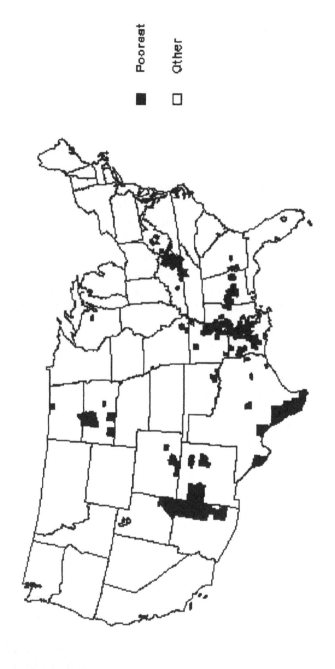

Poorest
Other

Figure 33. Spatial Distribution of the Poorest 5 Percent of Counties in the U.S. - 1990

variable used was the income index (INDEX). In order to make INDEX for 1980 and 1990 comparable, the 1980 INDEX was expressed in 1990 dollars using a conversion constant (U.S. Department of Commerce 1992b). Data observations for 1980 were assigned a dummy variable (YEAR) value of 0 while those for 1990 were assigned a value of 1.

The initial model of poverty expresses INDEX as a function of the Factors 1 through 9 generated by the previously conducted principal components analysis.

$$\text{LogINDEX} = k_0 + k_1 \text{ FACTOR1} + k_2 \text{ FACTOR2} \ldots +$$
$$k_n \text{ FACTORn} \qquad \text{(Equation 6.5)}$$

The parameters in the initial model (Equation 6.5) were expanded by the dummy variable YEAR and are expressed as follows:

$$k_i = k_{i0} + k_{i1} \text{YEAR}$$

$$\text{(Equation 6.6)}$$

where k_i refers to the model parameters in Equation 4.6.

$i = 0$ to n.

The series of expansion equations relate LogINDEX to the variation of each factor with time.

Substitution of the expansion equations into the initial model (Equation 6.5) yields a terminal model as follows:

$$\text{LogINDEX} = k_{00} + k_{01} \text{YEAR} + k_{10} \text{ FACTOR1} +$$
$$k_{11} \text{FACTOR1*YEAR} \ldots\ldots k_{n0} \text{ FACTORn} +$$
$$k_{n1} \text{FACTORn*YEAR} \qquad \text{(Equation 6.7)}$$

This terminal model will be used in the analysis of temporal variation in poverty from 1980 to 1990 in which terms in the terminal model that are cross products with the variable YEAR represent the variation in the parameter from 1980 to 1990. The terms that are not cross products give the parameters for 1980. Stepwise regression was used to select the "best" model that offers a maximum R^2 while keeping all terms statistically significant.

Results and Discussion

Location of Poverty Cores . Appendix A shows the 1990 ranking by the income index of counties who were poorest in 1980. These rankings reveal a relative stability within the poorest segment of counties over the decade of the 1980s. Of the 155 counties identified as the poorest 5 percent in 1980, 100 still remained in this segment in 1990. While some counties, such as Faulk, Miner and Sanborn Counties, South Dakota, did appear to make a meteoric rise out of poverty, most remained among the poorest counties in America. For those counties which remained within the most impoverished 5 percent there was some jockeying of relative position. Starr County, Texas took over the dubious honor of being the most impoverished county in America in 1990 from Tunica County, Mississippi. Tunica County, meanwhile rose seven positions in the ranking to become the eighth most impoverished county. Such changes do little to change the spatial picture of American poverty in 1990. Mapping the poorest 5 percent of counties in 1990 shows that these counties are even more spatially concentrated than they were in 1980. Coherent areas of extreme poverty clearly remain centered in South Dakota, eastern Arizona, the Texas-Mexican border, the Mississippi Delta, coastal plain Alabama, and the central Appalachians of eastern Kentucky (Figure 27).

Mapping both Per Capita Income and the income index for 1990 supports the view that the spatial distribution of American poverty remains essentially unchanged. The five distinct major cores of poverty identified to have existed in 1980 clearly persisted until 1990. However, some comments can be made on changes in poverty and its spatial distribution in the decade of the 1980s.

Figure 28 reveals that while similar cores of poverty have persisted since 1980, the spatial extent of the cores has undergone some change. The poverty area in South Dakota still persists but appears to have contracted just a little. This contraction is not typical of the other cores of poverty. The poverty core centered in northeastern Arizona remains as extensive as in 1980 and appears to have remained essentially spatially stable throughout the 1980s. Despite lack of confirmation of the Texas-Mexican border core in previous research (Brunn and Wheeler 1971), it appears that this area of poverty has persisted throughout the 1980s. In fact the impoverished area located along the Texas-Mexican border has expanded and has become linked by a band of poverty to the Mississippi Delta poverty region. The poverty area located in the Mississippi Delta region has also expanded,

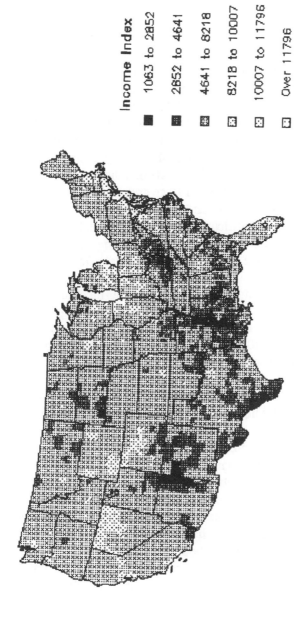

Figure 34. Spatial Variation in the Income Index in the U.S. - 1990

especially at its southern extreme in Louisiana. In 1990 the north-south axis of this poverty core stretched along both sides of the Mississippi River from central Arkansas to the Gulf coast. Although the band of poverty that spanned the southern coastal plain from South Carolina to the Delta poverty core got narrower during the 1980s, it still persisted. By 1990 the east-west axis had expanded to the west to include parts of east Texas that had been identified as poor in 1959 by Brunn and Wheeler (1971). The node of intense poverty at the conjunction of the two axes that form this core was still apparent in 1990. This node showed no indication of contraction during the 1980s. Indeed, poverty appears to be even more concentrated in this area. There has also been a clear expansion of the distinct area of poverty centered in the Appalachians of eastern Kentucky. It appears that the southern extent of this core has expanded and also that a large area of West Virginia has been encompassed. It also seems that in conjunction with an expansion of poverty has come an intensification of the poverty being experienced (Figure 28). This expansion and intensification of Appalachian poverty comes despite the extensive federal, state, and local efforts of the last thirty years.

Figure 29, which displays Per Capita Income in 1990, presents a somewhat different picture. While similar areas of poverty are identified, poverty appears to be a much less widely dispersed phenomenon than in 1980. Poverty areas revealed by mapping Per Capita Income are much less prominent and intense than those revealed by mapping the income index. The spatial pattern of Per Capita Income suggests progress is being made against poverty. The spatial pattern of the income index contradicts this conclusion and suggests poverty is expanding.

While the maps address the question of the spatial distribution of United States poverty and changes over the decade of the 1980s, they do not reveal if the poor are relatively better off or if the gap between rich and poor is changing. Table 9 presents a comparison of the 1980 and 1990 income index in terms of constant 1990 dollars and as a percentage of the average income index for the nation. Expressing the income index in 1990 dollars for both years reveals that the poorest counties in America have become poorer. These counties also moved further away from the average income index over the decade of the 1980s. In contrast, the most affluent counties have become more affluent. Expressed in constant 1990 dollars their income has

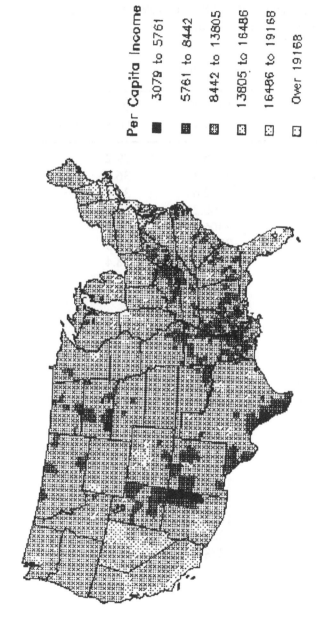

Figure 35. Spatial Variation in the Unadjusted Per Capita Income in the U.S. - 1990

175

Table 9. Comparison of the 1980 and 1990 Income Index for the Poorest and Most Affluent Counties.

1980

County	INDEX in 1990 $	INDEX as % of Average
Tunica, MS	1756.7	29.2
Hancock, TN	2174.4	36.1
Starr, TX	2192.6	36.4
Holmes, MS	2260.2	37.5
Greene, AL	2306.5	38.3
Costilla, CO	2310.3	38.3
Jefferson, MS	2316.5	38.4
Lee, AR	2372.4	39.4
Owsley, KY	2376.4	39.4
McCreary, KY	2432.2	40.4
Clay, KY	2472.9	41.0
Shannon, SD	2512.7	41.7
Jackson, KY	2518.4	41.8
Humphreys, MS	2532.2	42.0
East Carroll, LA	2535.6	42.1
.		
.		
.		
.		
Marin, CA	11984.4	198.9
Fairfax, VA	12403.3	205.8
Montgomery, MD	12831.7	213.0
Falls Church, VA	12879.3	213.7
Los Alamos, NM	12887.9	213.9

Table 9. Continued

	1990	
County	INDEX in 1990 $	INDEX as % of Average
Starr, TX	1581.5	24.6
Holmes,MS	1624.6	25.3
Owsley, KY	1674.8	26.0
East Carroll, LA	1758.8	27.4
Shannon, SD	1801.8	28.0
Apache, AZ	1977.2	30.8
Jefferson, MS	1990.4	31.0
Tunica, MS	2041.7	31.8
Zavala, TX	2066.5	32.1
Maverick, TX	2094.5	32.6
Breathitt, KY	2167.5	33.7
McCreary, KY	2176.5	33.9
West Feliciana, LA	2386.4	37.1
Mora, NM	2418.0	37.6
Humphreys, MS	2442.2	38.0
.		
.		
.		
.		
Howard, MD	14702.8	228.7
Montgomery, MD	14822.9	230.5
Fairfax, VA	15396.8	239.5
Falls Church, VA	16101.4	250.4
Los Alamos, NM	17448.5	271.4

substantially increased. In addition, this income is a greater percent of the average income index nationwide.

Along with a concentration of extreme poverty in the core areas identified, areas of below average income index have expanded. In 1980, 1607 or 51.7% of counties fell below the national average in terms of the income index. By 1990, 1693 counties or 54.4% commanded less than the average income index.

The general changes which have taken place in United States poverty from 1980 to 1990 can be summarized as follows. Poverty has become increasingly spatially concentrated. Poverty has also become more spatially extensive when viewed at a county level. In addition, income has increasingly polarized between rich and poor. The income commanded by the poorest counties has been reduced in real terms and is lower when compared to the national average. The most affluent counties, by contrast, enjoy substantially higher real income that has increased when compared to national averages. The 1980s, the decade of Reaganomics, has brought expected changes. Devine et al's (1992) contention that U.S. society has become increasingly unequal and polarized is supported by this research. The five poverty cores identified to exist in 1980 have persisted through the decade. Only one of the five poverty regions has shown any sign of contraction while the other four have expanded.

In addition to temporal variation in the spatial distribution of poverty, there is also possible variation in the impacts of the concomitants of poverty and characteristics of the population over the decade of the 1980s. To explore such temporal variations a Principal Components Analysis and subsequent Expansion regression modeling were conducted.

Principal Components Analysis. The Principal Components Analysis conducted on the merged combined 1980/1990 data set produced 9 factors with an eigenvalue over one (Table 10). These factors represent dimensions of poverty and characteristics of the population over the decade. The 9 factors incorporate over 87% of the information contained within the original data set. In addition to a high combined communality, the factors also represent all the original variables well. Table 11 provides communality estimates for each original variable; 13 are over 0.90 and the remaining two are over 0.82.

Interpretation of the factors generated involved assessment of the factor loadings. Table 12 presents the variables that were considered significantly loaded on each factor as well as the value of these

Table 10. Eigenvalues of the 1980/1990 PCA factors.

Factor	Rotated Eigenvalues	% Varience Explained
1	2.95	18.44
2	2.34	14.63
3	1.80	11.25
4	1.33	8.31
5	1.17	7.31
6	1.13	7.06
7	1.10	6.88
8	1.08	6.75
9	1.08	6.75
		87.38

Table 11. Communality Estimates for the Variables Used in the 1980/1990 Principal Components Analysis.

Variable	Communality
URBAN	.978
WHITE	.974
BLACK	.986
INDIAN	.987
SPANISH	.977
YOUNG	.850
TEENS	.828
OLD	.906
PERFEM	.905
PLUMB	.960
OVERCR	.911
HSCHOOL	.899
COLLEGE	.911
UNEMPL	.984
COMMUT	.987
LABFEM	.920

Table 12. Significant Factor Loadings and Interpreted Factor
Meanings, 1980/1990 Factors.

Factor		Loading	Meaning
1	BLACK	.957	Black - white duality.
	PERFEM	.822	
	WHITE	-.932	
2	YOUNG	.875	Age structure.
	TEENS	.827	
	OLD	-.697	
3	COLLEGE	.891	Education level.
	HSCHOOL	.813	
4	SPANISH	.971	Spanish culture and ethnicity.
5	INDIAN	.966	Native American culture and ethnicity.
6	PLUMB	.890	Housing conditions.
7	URBAN	.893	Urban residential status.
8	UNEMPL	.952	Employment status.
9	COMMUT	.976	Labor mobility.

loadings. The factor loadings, and hence interpretations, are relatively clear and uncomplicated.

Factor 1 is interpreted to encompass the black-white duality and is very similar to Factor 1 generated by the 1980 data alone. The variable BLACK is strongly positively loaded, while WHITE is strongly negatively loaded. The additional heavy loading on PERFEM is consistent with the interpretation, since the propensity of African American families to be female headed is well documented. Factor 1 is anticipated to be negatively associated with INDEX.

Factor 2 is strongly positively loaded on the variables YOUNG and TEENS and negatively loaded on OLD. Thus, this factor is said to indicate the age structure of a county's population. It differs from the 1980 extreme age population factor in that OLD and YOUNG have different signs. The expected relationship to poverty of this factor is difficult to assess since both the aged and the very young have been identified as at high risk of poverty. However, a consensus appears to suggest that the aged are no longer as likely to be poor as are young children because of the extensive social programs that benefit the elderly. Thus, it is expected that this factor will be positively related to poverty.

Factor 3 clearly represents educational level since both COLLEGE and HSCHOOL are strongly positively loaded. This factor is anticipated to be negatively related to poverty. This factor has no clear counterpart generated from the 1980 data; instead, in 1980 educational level was incorporated into the non-yuppieness factor.

Factor 4 is significantly loaded on one variable SPANISH, and is thus said to represent Spanish culture and ethnicity. This factor is very similar to Factor 8 used in the 1980 analysis. Spanish ethnicity is expected to be positively linked to poverty.

Factor 5 is similarly loaded on just one variable INDIAN, as was Factor 9 in the 1980 analysis. Factor 5 is interpreted to indicate Native American culture and ethnicity and is expected to be positively associated with impoverishment.

Factor 6's strongest positive loading is on PLUMB and in addition it is loaded moderately on OVERCR, hence this factor is said to be an indicator of housing conditions. Since PLUMB is a measure of housing that lacks complete plumbing facilities this factor is anticipated to be positively linked to poverty. No such factor resulted from the 1980 PCA. Instead, housing quality and urban residential status were closely tied in 1980, instead of showing as separate factors as they do in the combined 1980-1990 data set.

Factor 7, loaded positively on URBAN, is interpreted to indicate urban residential status. Urban residential status is expected to be associated with relative affluence and hence this factor is anticipated to have a negative relationship to poverty.

Factor 8 is strongly positively loaded on the variable UNEMPL and so represents employment status. Unemployment, and thus factor 8, is expected to be positively associated to poverty. This factor is similar to Factor 5 which resulted from the 1980 analysis.

The final factor, factor 9, indicates labor mobility, since it is positively loaded on the variable COMMUT. This factor is very similar to Factor 12 of the 1980 data analysis. It is anticipated that labor mobility will be negatively associated with poverty.

Temporal Variation in the Poverty Model, 1980 - 1990. The full regression model given by Equation 6.7 includes 19 variables, while the stepwise procedure indicates the "best" model includes 14 of these variables. The parameters for both the full and stepwise models are reported in Appendix B4.

The full model reveals that all the factors in the model were significant for 1980. A majority of the factors included in the model show shifts from 1980 to 1990, indicating that significant changes have taken place during the decade of the 1980s. However, not all factors show any significant shift over the ten years from 1980 to 1990. Factors 4, 6, and 7 show no statistically significant change over the decade and the temporal change in Factor 9 is relatively weak. The "best" model confirms the indications of the full model. The 1990 terms for Factors 4, 6, 7, and 9 are absent from the model. This indicates that the effect of Spanish culture and ethnicity, housing conditions, urban residential status, and labor mobility on the income index remained stable during the 1980s. Removal of these four terms from the model only slightly reduced the R^2 from 0.6967 to 0.6958.

The parameter estimates for 1980, given the restricted data set, broadly confirm the results presented earlier in the general model of poverty for the United States. Race and education level (contained within non-yuppieness in 1980) are both crucial to poverty status. In addition, poor housing conditions are associated with low levels of the income index. Also, both Spanish and Native American ethnicity are associated with low income but the relationship is less clear than for

African Americans. Unemployment is also linked to impoverishment but is overshadowed by the influence of the black-white racial duality. The picture of 1980 poverty remains one where race and education level are characteristics most strongly associated with income as expressed by INDEX. Relative affluence is associated with high levels of education and with urban residential status. The weakest factor in the full model is that of age structure. The problems with this factor have been mentioned earlier in that both age extremes may be associated with poverty and so the relationship of Factor 2 with INDEX is not clear.

The 1990 parameters presented in Table 13 indicate that important changes took place between 1980 and 1990 in the influence of certain concomitants of poverty and characteristics of the population on poverty. Only one "depressor" of income weakened during the 1980s. It appears that by 1990 Native American ethnicity (Factor 5) had become somewhat less likely to be associated with poverty. This result is consistent with the slight contraction of some poverty cores dominated by Native American population noted earlier.

In contrast, both the black-white duality (Factor 1) and unemployment (Factor 8) strengthened considerably in their impact during recent years. Perhaps the most important insight the temporal model provides is with respect to these two factors. While to be African American appears to be even more powerfully linked to poverty in 1990 than in 1980 this racial dimension of poverty was overshadowed by the importance of employment status during the decade. Employment status surged in importance as a predictor of income, so that by 1990 it had become more crucial than race.

The one strong elevator of income in 1980 was education level. By 1990 education level (Factor 3) had become even more powerfully positively linked to the income index. Thus, while race, and ethnicity, remained important to poverty in 1990, it was unemployment and lack of education that emerged as pivotal.

The remaining factor to exhibit significant temporal change from 1980 to 1990 was age structure (Factor 2). The influence of age structure remained weak in 1990 but was in the reverse direction than was exhibited in 1980. In 1980 the more dominated a county population was by the young, the more likely was that population to be poor. By 1990, it seems that the elderly had a stronger link to poverty than did the young.

The results of the temporally variant model indicate that the effect of several important factors has changed over the decade of the

Table 13. Recomputed 1980 and 1990 Parameters from Appendix B4 .

Variables	Parameter Estimates		
	1980	1990	Ratio 1990/1980
Intercept	8.6529	8.7363	1.0096
Black-White Duality	-0.0722	-0.0911	1.2618
Age Structure	-0.0088	0.0043	0.0045
Education Level	0.1148	0.1538	1.3397
Spanish Ethnicity	-0.0465	-0.0465	1.0
Native American Ethnicity	-0.0374	-0.0231	0.6176
Hounsing Conditions	0.0913	0.0913	1.0
Urban Status	0.0789	0.0789	1.0
Employment Status	-0.0387	-0.1123	2.9018
Labor Mobility	0.0550	0.0550	1.0

1980s. These changes can be summarized as follows. By 1990 Native Americans had made some gains in their poverty status. This result is consistent with the slight contraction of some poverty cores dominated by Native American population noted earlier. A visit to the southwestern poverty core in 1991 by the author gave little indication of extensive economic activity in this most extensive and concentrated area of Native American population. It is clear from the 1980 and 1990 maps (Figures 5 and 27) of the poorest 5 percent of counties, that these very poor counties had a greater representation in the southwestern core by 1990. These gains by Native Americans seem to have been made elsewhere, possibly in locations with a less intense concentration of this population. During the 1980s some Native Americans began to exploit a few economic possibilities. For example, some Native American groups opened gaming enterprises to use one of the few advantages they possess. A particularly successful example of economic activity organized and run by Native Americans can be found in Mississippi. The Choctaw reservation operates a large industrial park and a gaming operation is planned. These economic activities have brought jobs, increased affluence, and an enhancement of traditional culture.

In contrast, both the black-white duality and unemployment strengthened considerably in their impact during recent years. In 1980, race was a much more powerful indicator of poverty than unemployment; however, unemployment eclipsed race in its impact during the 1980s. This change appears to support researchers such as Wilson who point to the primacy of structural employment factors over race. During the 1980s the five poverty cores identified have experienced very different changes in unemployment. Table 14 illustrates these changes. The Appalachian core clearly suffered high unemployment in 1980. By 1990 most, but not all, counties had experienced a small decline in joblessness. The employment situation in Appalachia was bad in 1980 and remained so in 1990. The Texan border core showed great variation in county unemployment levels in 1980. By 1990 the situation in all counties had worsened considerably and all counties were suffering high unemployment levels. The Southwestern poverty core experienced very mixed changes in unemployment during the 1980s. In some counties unemployment levels remained virtually unchanged, in some levels declined, and in others there was a sharp increase in unemployment. The unemployment situation in the Dakotan poverty core worsened considerably from 1980 to 1990. Rates already high in

Table 14. Changes in Unemployment 1980 to 1990 for Selected
Poverty Counties.

| County/State | Unemployment Rate | |
	1980	1990
Appalachian Core		
Adair KY	7.2	5.7
Bath KY	13.3	9.6
Bell KY	15.0	14.1
Breathitt KY	12.9	15.8
Carter KY	14.0	11.6
Texan Border Core		
Maverick TX	13.5	21.9
Presidio TX	4.1	10.3
Starr TX	12.4	18.8
Webb TX	6.8	11.6
Zavala TX	12.7	19.7
Southwestern Core		
Apache AZ	12.6	23.6
Guadalupe NM	12.1	6.4
Mora NM	18.2	16.7
McKinley NM	11.4	13.6
San Miguel NM	12.4	12.1
Dakotan Core		
Buffalo SD	19.1	20.3
Corson SD	8.0	14.8
Shannon SD	19.3	30.5
Todd SD	11.4	20.6
Ziebach SD	6.1	15.6
Delta Core		
Holmes MS	10.2	15.8
Humpreys MS	10.9	7.4
Jefferson MS	16.5	25.5
Sharkey MS	7.5	10.1
Tunica MS	14.9	10.4

1980 in some cases became what can only be described as astronomical. Shannon County, South Dakota, for example had a 19.3% 1980 unemployment rate that had risen to 30.5% by 1990 (Table 14). The decade of the 1980s brought mixed employment results for Mississippi Delta poverty core counties. Unemployment rose in some Delta counties, fell in others, and remained stable in others. The changes that occurred in unemployment were very variable depending on the impoverished region under consideration. This seems to indicate that no consistent structural change, and thus no employment change, occurred that spanned all regions.

Within the context of the temporal study the strongest link to relative affluence in 1980 was education. In 1990 education remained the factor most powerfully associated with well-being and in fact during the decade this link strengthened. The ill educated and unemployed were more likely to be poor in 1990 than they were in 1980. The influence of race diminished somewhat, but this is not to deny that African Americans were still at a great disadvantage in 1990. The finding that unemployment was more strongly linked to poverty in 1990 than it was in 1980 has some possible theoretical implications. It may call into question the importance of the working poor and the thesis that an increasing number of jobs, especially in the service sector, place working people in poverty. It would seem that if this were a critical attribute of the 1980s, the link between unemployment and poverty would have weakened rather than strengthened.

From 1980 to 1990 there was some change in the age structure of those people most likely to be poor. The age-poverty link was never strong, largely because of the nature of the age structure factor. On balance, it seems that elderly Americans had an increasing likelihood to be poor by 1990 and surpassed the young as the most likely group to be impoverished. This may well be because many people over 65 years old had incomes just above the poverty line. Often, the elderly live on fixed incomes and any transfer payments tend not keep pace with rising costs of living. Thus, many of the elderly may have seen their relatively fixed incomes drop below a rising poverty line during the 1980s. This indicates that those who cite America's elderly as the success story of the war against poverty may have a misplaced optimism. It is evident that transfer payments without any structural change are likely to be only a temporary remedy for poverty. Social security payments will become increasingly difficult to sustain with the changing demographic structure of the population. Hence, the re-emergence of the elderly poor may be the beginning of a major problem for the 1990s and beyond.

VII

Conclusions and Policy Implications

CONCLUSIONS

Some unequivocal general conclusions emerge from the research presented here. The first research question posed was "What is the spatial distribution of the poor within the United States and can distinct poverty regions be identified?" The findings of this research showed several poverty cores are clearly identifiable. Cores of poverty are centered in Arizona-New Mexico, in the Texas-Mexico border area, in South Dakota, in the Mississippi Delta, and in central Appalachia. In these cores poverty is endemic. The geography of United States poverty is distinct and the spatiality of the phenomenon is an important dimension in the search for understanding and solutions.

The second research question explored was "What are the concomitants of income poverty, and the general characteristics of the poverty population?" Again some clear conclusions emerge. The racial, ethnic, age, employment status, educational, and mobility characteristics of the poverty population differ from those of America at large. Individuals who are African American, who are either very young or old, who are poorly educated, or who are unemployed are most likely to live in poverty. In addition, there are clear concomitants to income poverty, particularly in the areas of economic structure. An urban retail economy is particularly strongly associated with high income levels.

The third research question identified was "How do identified concomitants of income poverty and characteristics of the poverty population vary in importance spatially?" Indubitably, poverty is not monolithic. Rather, substantial variations exist both regionally and between urban and rural counties. In rural America poverty is less strongly linked to employment than in urban counties. In addition, a

189

manufacturing economy, while it benefits urban areas, does little to raise income in rural counties. Regional variations result in distinct profiles for the identified poverty cores. In the Southwestern core lack of education and the limited manufacturing economy is most strongly associated with poverty. In the Border poverty core unemployment is overwhelmingly the most important factor in high levels of poverty. The Dakotan poverty core suffers from a multidimensional lack of economic activity. In the Delta/Southern poverty core, poverty is strongly associated with the African American population. Appalachian poverty appears to be deeply entrenched and multifaceted.

The final research question posed was "Have the spatial distribution of poverty, the concomitants of income poverty, and characteristics of the poverty population changed from 1980 to 1990?" While significant temporal variations are limited they are important. During the 1980s there has been a crucial shift in the factors most associated with poverty in the United States. Education level became more important to well-being from 1980 to 1990, while unemployment and the black-white duality became more strongly associated with poverty. However, while both unemployment and the black-white duality became more strongly linked to poverty, unemployment had overshadowed race as the most powerful factor by 1990.

The research presented here deepens understanding of existent theory and thus, makes an important theoretical contribution. It is clear that theory cannot be ubiquitously applied spatially or temporally. This study highlights the fact that theoretical explanations and concomitants of poverty are geographically and temporally variant. Theory that is pivotal in explaining poverty at one place or at one time may be less relevant at others. Poverty is not monolithic and it seems reasonable to suggest that theoretical explanations should also be spatially and temporary flexible. Thus, the element of individual responsibility cannot be summarily discounted in favor of structural explanation, nor the economic system seen as necessarily primary over social factors. It is necessary to view place as perhaps a unique conjunction of explanation as well as circumstance. Some clear examples have emerged within this research. Theory that places responsibility directly on the operation of capitalism is clearly primary within Appalachia. In the Delta/Southern poverty core discrimination, both institutionalized and individual, that is intimately linked to capitalism in the South, is theoretically central. Even where one broader theoretical explanation appears applicable in two regions, the problems and hence policy implications to emerge can be very different.

POLICY IMPLICATIONS
AND THEORETICAL INSIGHTS

The first policy implication to come from this research is that any anti-poverty program should contain a spatial dimension. Poverty is concentrated in several cores, and while this does not mean that general national level policies will be totally ineffective, it does indicate targeting of specific areas may be necessary. For example signing universal health care coverage into law does mean that theoretically every American would be entitled to health care. However, it would do nothing to address the lack of physicians, clinics, and hospitals that are experienced in many rural areas. Unless these problems that pertain to specific locales are addressed, the reality of lack of health care would remain unchanged.

The general characteristics of the poverty population suggest that social programs should be targeted toward specific groups. Clearly, race and ethnicity remain major issues. It seems that equal opportunity for all Americans has not been achieved and is perhaps an ideal that should be strived for within policy. In addition, both the old and the young remain vulnerable groups. Children and the elderly who live in poverty need special help. One possibility that would help the elderly is to replace the present social security system with one that draws contributions from all but which makes payments based on the level of need. It seems to make little sense to make transfer payments to the wealthy when the money is desperately needed elsewhere. Of course, the attitude of people that having paid into the system they have a right to be paid from it remains a major barrier to implementation of such a policy.

A national health service that offers comprehensive health coverage to all would be an enormous help to all the poor. The current system which offers the best care in the world to those who can afford to pay for it and no care to many who cannot, reinforces a dual society. This system should offer a full range of reproductive services. The added burden of an unwanted child is obviously one that poor families are least able to afford. Poor families, including many who rely on welfare payments, would benefit enormously from an affordable and safe system of childcare. Many conservative critics of the welfare system characterize the poor as lazy and unwilling to work. Often the reality is that taking a job means paying high childcare costs. Costs are often so high that income is inadequate to survive. In addition to the expense, childcare is poorly regulated and children placed in physical or emotional jeopardy. Instead of withdrawing all welfare support from people who do find jobs, the state should provide daycare or pay its costs for those who cannot.

Regional variations in the nature of poverty and characteristics of the poverty population have important policy implications for poverty cores. It is clear that one policy or policy package does not offer appropriate solutions to all areas where the poor are concentrated. Policy must address the regionally variable nature and causes of poverty and must be sensitive to cultural needs and expressed wishes of the intended beneficiaries. It should be remembered that the policy suggestions offered, although linked to the results obtained from the regionally variant model, do not emanate directly from it. The regional model offers a view of variation by geographic division, while the poverty cores span a much smaller area. Each poverty core has a distinct cultural, social, political, and economic character and this spatial context must be a major concern when offering policy suggestions.

The Southwestern Core

The Southwestern poverty core centered in Arizona-New Mexico is located in the Mountain geographic division. It is in this division that non-yuppieness and lack of manufacturing activity has a particularly strong influence, and employment status a weak influence on poverty status. Thus, the results of this research suggest job creation programs might be ineffective. Instead, alleviation of poverty would be best accomplished by efforts to improve education and stimulate manufacturing and urban growth. The nature of manufacturing should be a concern. The southern experience shows that attracting low wage, labor intensive industries will do little to alleviate poverty. However, pursuit of a policy to stimulate manufacturing and urban growth, although indicated as desirable by this research for the Mountain geographic division, may not be the most appropriate policy for the southeastern poverty core. Even in the unlikely event that urban growth and appropriate manufacturing could be stimulated, such policy would ignore the needs of Native Americans as a distinct cultural group. Central to these needs are preservation of culture and basic elements of traditional lifestyles. However, this is not to say that the growth of some urban centers and types of manufacturing would not be helpful. The southwestern poverty core seems to have fewer direct economic links with the capitalist urban industrial system than some other regions. While this has resulted in poverty, it may also be an advantage for the future. The core does not have to take an initial step to extricate itself from extensive negative capitalistic connections. To preserve their culture and make economic gains it seems that only residents themselves are qualified to identify acceptable goals and appropriate action.

Some progress has been made recently as Native Americans have become politically empowered and have sought settlement of land claims that have been long ignored. In addition, Native Americans have begun to organize in order to reap more benefit from the lands they already have. It is this kind of action by Native Americans themselves that seems to offer potential for economic progress while at the same time strengthening cultural cohesion. The Navajo nation, the largest Native American group in the southwestern poverty core, have become active leaders in a drive to end exploitation of Native Americans and of tribal lands (Guinness and Bradshaw 1985).

Existing institutions appear to be inadequate and have failed to address deep social problems. Future policy designed by outside agencies should seek to support and expand efforts, rather than to impose mainstream cultural ideas of development. Part of this support, this research indicates, must be in the form of adequate health care provision that addresses the health needs of residents. A few widely separated hospitals and an overburdened system of clinics is not enough. One suggestion for improvement might be community health cooperatives. Training individuals who are part of the community to deliver some basic health care services might have far reaching benefits. Language and cultural barriers in this way can be largely overcome and traditional beliefs woven into provision of modern medical care. The training required to gain competency in delivering vaccinations, in judgment of when to prescribe some basic medications, in recognition of common health problems, or in recognizing prenatal problems is not extensive. In addition, community based programs to educate, provide treatment, and offer alternatives to alcohol abuse are more likely to be effective than are dictates and programs from outside.

This research also indicates that education is another area in which support from the outside might be beneficial. Producing well educated Native Americans capable of pursuing the communities' legal and political interests would be an important step forward. Again, production line education without regard to the special cultural and life needs of Native Americans is likely to be viewed as destructive and irrelevant. There is no reason why basic literacy and even higher levels of achievement cannot be placed in what is a rich cultural heritage. Native Americans do command both natural and cultural resources. The key seems to be empowerment and mobilization of the community to use the resources available. Although Gaventa (1980) wrote specifically about the Appalachian region some of his comments concerning power and powerlessness may be usefully applied to the Native American lands of the southwest. Visiting the area and talking to residents gives a sense of apathy and powerlessness, as well as a hopeless kind of anger about the losses of the past and reality of the present. There seemed to be limited willingness to discuss the future and certainly no vision for long range change. Gaventa's observations of the role of history in the development of powerlessness seem particularly relevant. In addition, his point is well made that powerlessness is not an individually rooted problem and solutions are not be found at the individual level.

The Texas Border Core

        In the Texas border poverty core job creation is crucial to reduction of poverty. The negative effects of unemployment overwhelm all others in this area, and so this study indicates that one focus of policy should be the provision of jobs. While unemployment was not especially high in this poverty core in 1980, the power of unemployment to depress income suggests that employment status cannot be ignored. In 1980 the southern extent of the border poverty core levels of unemployment were high, indicating that here, especially, employment status was a real problem; by 1990 this whole core was engulfed in the problem of high unemployment. One solution for the area's problem is government action to stimulate job growth. It seems likely that a lot of pressure on the region comes from illegal movement of people from Mexico, as well as a concentration of manufacturing which uses cheap labor within Mexico. Thus, part of policy to improve life in this border core must focus on a tandem improvement in Mexico. Not only would this be the most effective recourse against illegal migration but it would also expand the market for consumer goods and address the employment vacuum on the U.S. side of the border. A market for goods in Mexico might make this border area an attractive location for new economic activity that could provide much needed employment. In order for this to occur, wages and benefits for Mexican workers must be made to approach levels in the United States so as to eliminate the benefits gained by employers by their exploitation. With the development of Mexico's economy, the border core's cultural and linguistic integration with Mexico may become a perceived asset rather than a negative characteristic for outside investors. The recent passing of the North American Free Trade Agreement (NAFTA) may offer some hope for economic growth in this border region, both on the U.S. and Mexican sides. NAFTA presents both a danger and an opportunity. A danger exists that communities will, out of desperation to attract jobs, offer concessions that have a high cost for limited benefit. The region may be drawn even more negatively into capitalism than it is already. Exploitation of the border region as a cheap wage interface with Mexico is not likely to bring benefit to a majority of residents. As any development associated with NAFTA proceeds the local population may have some opportunities to direct the process. Such control by the community rather than a local elite will not be easy. Perhaps organization of all border counties into a single policy unit is an option that would minimize competition between counties. Such a regional unit would present a united front to potential investors interested in tapping NAFTA's possibilities and limit the use of space for profit to a few.

The border poverty core has a strong ethnically Spanish character. Policy aimed at such a coherent cultural region must be sensitive to the expressed needs and desires of the population which may have their own ideas about what constitutes desirable change. It seems that a strategy to combat poverty in the border core might be most efficient if targeted toward the ethnically Spanish population, especially in the area of job provision. However, such a strategy, if initiated outside the region, is likely to be most effective if formulated and coordinated under close consultation with the population it is designed to serve.

The Dakotan Core

The poverty core identified as being centered in South Dakota appears likely to benefit most from a multidimensional expansion of the economy. Expansion of the urban retail, manufacturing, and health service economies could all benefit the region. This economic expansion seems especially critical in light of the explosive rise in unemployment during the 1980s in this region. As a part of this expansion it seems reasonable to suggest improvement of the transportation network. Both an internal network and external links are necessary for the kind of multidimensional economic expansion this region appears to need. An extensive improvement in the road network cannot be completely achieved through local or even state action in isolation. While impetus may come from the local level, state financing must be supplemented by federal dollars.

It has also been indicated that large numbers of young children are a financial drain on families in this poverty core. Therefore, programs that provide services for these children seem appropriate. These might include pediatric health care, early childhood education, and childcare programs. As suggested for the Southwestern poverty core, community health cooperatives might be helpful in providing some basic health care services. Similar comments concerning improvements in education can be made for the Dakotan as for the southwestern core, since both show remarkable spatial conversion with Native American reservation lands. Again, this core does have resources. For example, Shannon County, where unemployment was an horrendous 30.5% in 1990 is the location of spectacular badlands scenery and historic sites such as Wounded Knee. Tourism in the region could be controlled and developed by the Native American population in order to maximize revenues and minimize disruption.

The Delta/Southern Core

What unifies the Delta poverty core is its history of slavery and current large African American population. This study indicates the importance of all the racial and ethnic factors, a fact that seems to point to systematic discrimination against non-whites. It is theories that point to discrimination as being at the root of poverty that seem most applicable in the Delta poverty core. Thus, it is discrimination at all levels that must be addressed. This a difficult task. At the individual level one cannot mandate the attitudes that prevail on both sides of the black-white racial division. These personal prejudices extend to such areas as employment and become part of social and economic structure. These structural discriminations are often subtle and difficult to prove. This makes them difficult to address, especially when many people, including those in government, appear to tolerate or even condone the situation. Only in cases where discrimination is particularly blatant can legal action be effective. In addition, legal recourse is very difficult for many victims of discrimination to initiate and follow through on. The poor lack finances, education, knowledge of discrimination law and legal recourses, and the resources necessary to overcome numerous roadblocks to action. It is within institutionalized discrimination that the most effective change can be made. Change within education would provide a good beginning. Many problems within education that lead to different outcomes for black and white students stem from the local nature of funding and control. Ideally, education would benefit from integrated national control, funding, and standards. Given the strong sentiment in the United States in support of local and state power, this suggestion is not practicable. A step in the right direction would be a state education system. Control must be taken out of the hands of local school boards and local components of funding must be discontinued. Even this change to a state school system will face fierce opposition especially from those who benefit from the current situation. These beneficiaries are powerful, and range from highly paid school superintendents to a host of other local elites and capitalists.

The Delta core also suffers from particularly adverse effects of unemployment. Thus, job creation efforts may be helpful. The nature and quality of jobs is an issue. The types of manufacturing jobs likely to be attracted may offer, at best, minimal benefit. There is some indication that stimulation of a more extensive urban retail economy would be beneficial. The impetus for such an expansion must come in large measure from local elites, including businessmen and politicians. With such a large proportion of the population deeply impoverished, the growth of an urban retail economy will be difficult to initiate and sustain. Fundamental to alleviation of poverty is a redistribution of wealth which is extremely polarized (Shaw 1990b). This polarization has deep historical roots that will be very difficult to modify. Agriculture, which is the backbone of the economy in the Delta, is characterized by large mechanized farms owned by a few individuals or companies. There is little demand for the unskilled labor of the majority of the population (Ford 1973). While this condition persists and just a tiny fraction of the population command the majority of wealth and income, there can be little progress against poverty in this region through expansion of the urban retail economy.

The Appalachian Core

Although the Appalachian poverty core is within the East South Central geographic division it is clear that the black-white duality is not the issue here. The fundamental problem is this region's place within American capitalism (Moore 1994; Pearson 1994). This historic and continuing role within capitalism has resulted in a multitude of social, economic, and political symptoms. It will be necessary to change both the economic role of the region as well as addressing the aforementioned symptoms that have taken on somewhat of a life of their own.

Diversification of the economy is necessary to move the region away from dependence on mining and forestry. A concerted effort has been underway to achieve this since the establishment of the Appalachian Regional Commission (ARC) in 1965. The Commission combined state and federal efforts and established Local Development Districts (LDDs) to link local areas to the state. Investments were made in infrastructure, public facilities and utilities, education, housing, health, public facilities, environmental planning and planned resource use. Focus was on identified growth centers within each LDD (Appalachian Regional Commission 1966a; 1966b; 1972; 1977; Bradshaw 1992). The result of these efforts has brought some benefit. Living conditions, education, and health have improved and the economy has diversified somewhat. However, the nature of that diversification is problematic and seems to have enmeshed the region more deeply in some of the most negative facets of American capitalism. New industries have been concentrated in such things as textiles and furniture which are labor intensive and low wage (Hamm 1973). Development of small business has been hampered by tight control of funds by banks within the region (Shaw 1996). These banks are small, locally owned and often only make personal loans and only to people they know well (Checchi and Company 1969). One possible solution to the problems of diversification might be to open the region to outside banking and break the monopoly of local banks. In addition, diversification should be encouraged outside designated growth centers as poor rural areas seem to reap little benefit and remain entrenched in poverty (Newman 1972).

Particularly destructive and pervasive symptoms of the region's historic and continuing role within capitalism are to be found within local politics. Substantial county government machinery exists even in financially strapped counties. Systems of county government include useless offices and idle officials. However, this misuse of resources is not the main problem. Positions as county officials such as the sheriff, county judge, and school board members may be bought. Individuals seek to buy these political positions because they hold extraordinary power that extends into the economic sphere (Wager 1962; Schrag 1972; Vance 1972). These local officials have the power to assign jobs and who gets them is often decided by political loyalties or family ties. This has particularly serious consequences for education. The whole school system is seen as corrupt and people who work within it are there for reasons other than merit. The result is poor quality education and many high school graduates that are illiterate (Duncan 1986). These local political figures are part of a local elite with links to the large coal and timber companies that dominate this region's economy (Schrag 1972; Wells 1977). As Gaventa (1980) observes, the power structure in Appalachia is well developed and will not be easy to change. Communities must not only recognize corruption but also organize and act to confront it. However, it seems that while corruption is recognized, quiescence, a product of a sense of powerlessness, is widespread (Gavanta 1980). One possible change, if it can be effected, involves the consolidation of counties. Another important goal is the removal of education from local control, and thus from its domination by the interests of elite groups and large corporations. In both these changes the State must take an initiative and play an active role.

Another important result of the alliance between the local elite and the large coal and timber companies is that very low taxes are paid by these companies (Osborne 1986). Companies evade honest reporting of ownership and mineral wealth (Millstone 1972), and even when taxes are paid they are miniscule. Szakos (1986) provides the example of Martin County, Kentucky where 81,333 acres of coal yielded an annual tax of $76. Large profits are made and wealth siphoned from the area but little money enters the local economy through taxes (Good 1972). The environmental costs of mining and timber exploitation far outweigh taxes paid (Brooks 1972; Armstrong-Cummings 1986). Coal and timber companies must be forced to pay taxes on the resources they exploit and the profits they make, that go to the area that suffers the costs. In addition, the costs of environmental damage should be paid by these companies. Finally industry, especially those attracted by financial incentives, should be assessed a severance tax should it decide to abandon the communities which have sponsored relocating firms. Given the local power structure these changes necessitate involvement of both the federal and state political machines. While capitalistic exploitation of the region remains unchecked, central Appalachia will continue to be a region of rich resources and an impoverished people.

FUTURE RESEARCH

Since explorations of the regional variation in the major concomitants of income poverty and characteristics of the poor presented are for 1980, one avenue for future research is to update these findings using 1990 census data. This should be possible very shortly as census data become fully available. It will be interesting to discover if the general regional variations presented for 1980 have persisted through the 1980s. It will also be important for future policy to explore the temporal variation of important factors for each poverty core. For example, unemployment clearly has become more crucial than race at the national level, but this may not be the case in important poverty regions such as the Delta.

This research has also pointed to the need for deeper understandings of the nature of place. As researchers such as Cooke (Cooke 1987a; 1987b; 1989), Storper (1987), Beauregard (1988), and Lovering (1989) contend, understanding of place must not be circumscribed by the limits or frameworks of existing theory. Beyond the application of universal theories such as Marxism and their articulation with prexisting systems and inequalities such as patriarchy and patronage, empirical research may be necessary to broaden understanding. In such circumstance an empirical focus does not seek to deny or tear down existing theory, but rather attempts to explain realities not well explained by existing theory. Empirical investigation is the first step in formulating supplemental theoretical insights that may be important in understanding spatial inequalities and the uniqueness of place. Thus, the study of the local is not a move from theory, but instead is an attempt to broaden and deepen theoretical understanding. Future research might usefully include a detailed study of the economic, social, political and cultural character of identified poverty regions. Each core of poverty must be viewed in terms of its history, internal dynamics, its regional position, and links to the broader economy.

Appendix A

The Poorest Five Percent of Counties in the Conterminous United States by Adjusted Per Capita Income (1980), and their Unadjusted Per Capita Income Ranking (lowest to highest) for 1980 and Adjusted Per Capita Income Ranking for 1990.

County-State	Adjusted PCI 1980	Rank 1980	Unadjusted PCI Rank 1980	Adjusted PCI Rank 1990
Tunica, MS	1107.60	1	167	8
Hancock, TN	1379.99	2	25	58
Starr, TX	1382.49	3	6	1
Holmes, MS	1425.08	4	56	2
Greene, AL	1454.26	5	210	18
Costilla, CO	1456.67	6	1712	72
Jefferson, MS	1460.61	7	466	7
Lee, AR	1495.82	8	215	25
Owsley, KY	1498.34	9	33	3
McCreary, KY	1533.54	10	18	12
Clay, KY	1559.21	11	254	42
Shannon, SD	1584.30	12	3	5
Jackson, KY	1587.88	13	21	49
Humphreys, MS	1596.62	14	177	15
East Carroll, LA	1598.75	15	204	4
Maverick, TX	1600.49	16	4	10
Sharkey, MS	1600.57	17	395	27
Lowndes, AL	1617.75	18	237	128
Wilcox, AL	1626.49	19	199	28
Chicot, AR	1651.01	20	285	30
Perry, AL	1676.43	21	77	65
Clinton, KY	1687.41	22	14	48

Morgan, KY	1696.66	23	47	69
Tallahatchie, MS	1697.53	24	86	71
Clay, GA	1702.38	25	262	95
Sanborn, SD	1702.56	26	795	1271
Hale, AL	1703.14	27	161	92
Ziebach, SD	1704.68	28	150	55
Zavala, TX	1712.37	29	349	9
Buffalo, SD	1714.89	30	147	50
Quitman, GA	1718.69	31	24	404
Quitman, MS	1718.69	31	282	67
Breathitt, KY	1721.38	33	186	11
Presidio, TX	1726.62	34	1649	21
Corson, SD	1764.96	35	110	76
Todd, SD	1767.96	36	64	20
Bullock, AL	1792.83	37	721	130
West Carroll, LA	1808.77	38	75	141
Apache, AZ	1811.49	39	122	6
Coahoma, MS	1824.04	40	584	38
Conejos, CO	1826.89	41	35	70
Phillips, AR	1829.90	42	400	35
Bolivar, MS	1850.00	43	241	37
Knox, KY	1871.76	44	84	22
Magoffin, KY	1879.94	45	157	26
Wolfe, KY	1880.28	46	48	23
Webb, TX	1894.99	47	297	34
Madison, LA	1901.32	48	32	41
Newton, AR	1918.66	49	8	222
Lee, KY	1924.40	50	61	62
Stewart, GA	1929.03	51	216	148
Mora, NM	1930.03	52	5	14
Macon, AL	1932.76	53	160	113
Searcy, AR	1940.00	54	46	132
Issaquena, MS	1954.71	55	9	39
Wheeler, GA	1956.12	56	104	279
Rockcastle, KY	1956.46	57	44	164
La Salle, TX	1963.32	58	225	127
St. Helena, LA	1970.31	59	139	110
Noxubee, MS	1971.34	60	79	87
Lawrence, KY	1975.75	61	294	100
Mellette, SD	1978.88	62	115	89

Sumter, AL	1982.37	63	211	19
Bennett, SD	1984.86	64	332	165
Jenkins, GA	1984.95	65	414	283
Casey, KY	1986.76	66	39	214
Monroe, AR	1991.17	67	771	85
Claiborne, TN	2003.78	68	183	308
Leflore, MS	2011.22	69	654	80
Brooks, GA	2018.27	70	386	494
Tensas, LA	2021.18	71	284	29
Wayne, KY	2024.48	72	59	59
Fentress, TN	2033.33	73	19	143
Wilkinson, MS	2037.30	74	374	24
Oregon, MO	2041.92	75	65	185
Sunflower, MS	2049.85	76	243	36
McKinley, NM	2063.27	77	308	33
Dimmit, TX	2063.69	78	248	17
San Miguel, NM	2068.51	79	89	211
Elliott, KY	2070.42	80	34	84
Bronx, NY	2076.32	81	2051	149
Miner, SD	2079.17	82	1217	1077
St. Francis, AR	2083.17	83	827	88
Guadalupe, NM	2084.67	84	63	46
Panola, MS	2085.71	85	242	258
Ripley, MO	2101.12	86	27	142
Willacy, TX	2105.74	87	152	45
Arthur, NE	2110.04	88	464	1799
Randolph, GA	2111.82	89	198	107
Douglas, SD	2113.24	90	587	335
Bell, KY	2122.90	91	831	32
Stone, AR	2124.54	92	58	350
Leslie, KY	2126.46	93	80	53
Menifee, KY	2129.47	94	17	150
Hancock, GA	2132.63	95	148	309
Hidalgo, TX	2141.24	96	149	43
Grant, ND	2142.76	97	1195	569
Tattnall, GA	2147.18	98	270	511
Faulk, SD	2154.04	99	1685	1080
Dewey, SD	2155.17	100	696	83
Yazoo, MS	2158.31	101	938	63
McPherson, SD	2165.13	102	1317	340

Webster, GA	2169.68	103	1493	452
Bertie, NC	2175.66	104	549	446
Kemper, MS	2176.60	105	52	126
Atkinson, GA	2183.93	106	362	472
Allendale, SC	2185.06	107	141	82
Whitley, KY	2185.31	108	592	93
Metcalfe, KY	2190.97	109	50	156
Charles Mix, SD	2204.44	110	753	277
Cumberland, KY	2204.81	111	109	105
Wayne, MO	2208.43	112	11	138
Covington, MS	2215.69	113	424	373
Sioux, ND	2216.27	114	828	81
Adair, OK	2216.78	115	69	342
Haywood, TN	2218.39	116	425	351
Johnston, OK	2224.14	117	112	271
Jefferson, FL	2224.34	118	296	1144
Val Verde, TX	2225.15	119	525	119
Gregory, SD	2231.46	120	1048	509
Jefferson, GA	2235.41	121	444	257
Scott, TN	2236.25	122	138	155
Choctaw, AL	2247.59	123	470	199
Claiborne, MS	2247.83	124	387	16
Russell, KY	2251.45	125	102	570
Richland, LA	2257.33	126	619	115
Attala, MS	2257.49	127	156	196
Conecuh, AL	2257.59	128	287	262
Woodruff, AR	2265.83	129	1566	122
San Juan, UT	2267.33	130	74	75
Webster, WV	2267.47	131	67	56
Amite, MS	2268.30	132	176	151
Pemiscot, MO	2274.55	133	379	97
Lafayette, AR	2276.60	134	441	86
Carroll, MS	2277.70	135	10	321
Lewis, KY	2278.64	136	73	186
Lake, MI	2282.21	137	443	168
Hyde, NC	2285.12	138	856	475
Douglas, MO	2287.27	139	37	425
Hudspeth, TX	2291.52	140	2720	223
Clay, WV	2292.83	141	143	57
Crenshaw, AL	2301.44	142	247	454

Franklin, FL	2303.30	143	76	268
Early, GA	2303.62	144	554	203
Catahoula, LA	2310.10	145	234	108
Leake, MS	2310.27	146	392	247
Atoka, OK	2310.62	147	49	102
Evangeline, LA	2311.90	148	447	60
Monroe, KY	2318.77	149	174	293
Rolette, ND	2324.42	150	860	136
Madison, ID	2330.91	151	233	194
Frio, TX	2335.01	152	227	52
Macon, GA	2338.34	153	473	231
Clay, TN	2338.35	154	95	785
Dooly, GA	2339.27	155	1202	153

Appendix B

1. Regression Statistics for Equation 5.1.

Independent Variables		Parameter Estimate	T Value	Probability > |T|
Intercept	k_0	8.2072	3133.81	.0001
Factor1	k_1	-0.1096	-41.85	.0001
Factor2	k_2	-0.1131	-43.18	.0001
Factor3	k_3	-0.0099	-3.79	.0002
Factor4	k_4	0.0939	35.84	.0001
Factor5	k_5	-0.1059	-40.44	.0001
Factor6	k_6	0.0342	13.08	.0001
Factor7	k_7	0.0441	16.84	.0001
Factor8	k_8	-0.0372	-14.20	.0001
Factor9	k_9	-0.0226	-8.65	.0001
Factor10	k_{10}	0.0091	3.47	.0005
Factor11	k_{11}	0.0085	3.25	.0012
Factor12	k_{12}	0.0259	9.88	.0001

F Value	616.724
F Probability	.0001
R^2	.7053
Adjusted R2	.7042

2. Regression Statistics for the Urban-Rural Model, Equation 5.3

Independent Variables/ Term		Parameter Estimate - Full Model	Parameter Estimate - Stepwise Model	F
Intercept	k_{00}	8.2010 **	8.2018	
TYPE	k_{01}	0.0582 **	0.0609 **	32.7
Factor1	k_{10}	-0.0152 **	-0.1049 **	1280.7
Factor1*TYPE	k_{11}	-0.0217 **	-0.0221 **	11.9
Factor2	k_{20}	-0.1230 **	-0.1233 **	1016.0
Factor2*TYPE	k_{21}	0.0277 **	0.0284 **	21.2
Factor3	k_{30}	-0.0179 **	-0.0166 **	38.6
Factor3*TYPE	k_{31}	0.0049	___	___
Factor4	k_{40}	0.0947 **	0.0950 **	679.7
Factor4*TYPE	k_{41}	-0.0355 **	-0.0383 **	21.3
Factor5	k_{50}	-0.0959 **	-0.0961 **	1100.2
Factor5*TYPE	k_{51}	-0.0318 **	-0.0308 **	22.6
Factor6	k_{60}	0.0160 **	0.0163 **	30.3
Factor6*TYPE	k_{61}	0.0476 **	0.0482 **	59.3
Factor7	k_{70}	0.0311 **	0.0356 **	173.0
Factor7*TYPE	k_{71}	0.0079	___	___
Factor8	k_{80}	-0.0337 **	-0.0329 **	61.3
Factor8*TYPE	k_{81}	-0.0337 *	-0.0143 **	6.7
Factor9	k_{90}	-0.0184 **	-0.0186 **	50.2
Factor9*TYPE	k_{91}	-0.0493 **	-0.0532 **	24.0

Factor10	$k_{10,0}$	0.0164 **	0.0161 **	22.8
Factor10*TYPE	$k_{10,1}$	-0.0196 **	-0.0205 **	14.5
Factor11	$k_{11,0}$	0.0085 **	0.0079 **	9.1
Factor11*TYPE	$k_{11,1}$	-0.0030	____	____
Factor12	$k_{12,0}$	0.0383 **	0.0380 **	139.1
Factor12*TYPE	$k_{12,1}$	-0.0191 **	-0.0183 **	11.1

F Value	330.38	375.32
F Probability	.0001	.0001

R^2	.7284	.7281
Adjusted R^2	.7262	

**	Significant at the .01 level
*	Significant at the .05 level

3. Regression Statistics for the Regional Model, Equation 5.5

Variables		Parameter Estimate - Full Model	Parameter Estimate - Stepwise Model	F
Intercept	k_{00}	8.2375 **	8.2122 **	
D1	k_{01}	0.1936 **	0.1792 **	103.1
D2	k_{02}	-0.0358	————	————
D3	k_{03}	-0.0561 **	————	————
D4	k_{04}	-0.0643 **	-0.0414 **	14.6
D5	k_{05}	-0.0972 **	-0.0633 **	46.4
D6	k_{06}	0.0694 **	0.1082 **	39.0
D7	k_{07}	-0.0442	————	————
D8	k_{08}	-0.0886	-0.0831 **	25.3
Factor1	k_{10}	-0.0128 **	-0.1151 **	1034.3
Factor1*D1	k_{11}	0.0691	————	————
Factor1*D2	k_{12}	-0.0339	————	————
Factor1*D3	k_{13}	0.0346 **	————	————
Factor1*D4	k_{14}	-0.0370 **	-0.0505 **	29.4
Factor1*D5	k_{15}	-0.0165	————	————
Factor1*D6	k_{16}	0.0781 **	0.0762 **	22.3
Factor1*D7	k_{17}	-0.0135	————	————
Factor1*D8	k_{18}	-0.0494	————	————
Factor2	k_{20}	-0.0743 **	-0.0976 **	1349.3
Factor2*D1	k_{21}	0.0115	————	————
Factor2*D2	k_{22}	-0.0540 **	-0.0305 **	14.3
Factor2*D3	k_{23}	-0.0213 *	————	————
Factor2*D4	k_{24}	-0.0271	————	————
Factor2*D5	k_{25}	-0.0320 **	————	————
Factor2*D6	k_{26}	-0.0307 **	————	————
Factor2*D7	k_{27}	-0.0162	————	————
Factor2*D8	k_{28}	0.0174	————	————

Factor3	k_{30}	-0.0290 **	-0.0293 **	68.6
Factor3*D1	k_{31}	0.0241	——	——
Factor3*D2	k_{32}	0.0207 **	0.0265 **	15.9
Factor3*D3	k_{33}	0.0291	0.0274 **	17.1
Factor3*D4	k_{34}	0.0134	——	——
Factor3*D5	k_{35}	0.0006	——	——
Factor3*D6	k_{36}	0.0453 **	0.0509 **	18.1
Factor3*D7	k_{37}	-0.0342	——	——
Factor3*D8	k_{38}	-0.0505	——	——
Factor4	k_{40}	0.0774 **	0.0828 **	538.6
Factor4*D1	k_{41}	-0.0604 **	-0.0544 **	12.2
Factor4*D2	k_{42}	-0.0241 *	-0.0341 **	17.4
Factor4*D3	k_{43}	-0.0156	-0.0161 *	4.6
Factor4*D4	k_{44}	0.0106	——	——
Factor4*D5	k_{45}	0.0353 **	0.0277 **	17.6
Factor4*D6	k_{46}	0.0191	——	——
Factor4*D7	k_{47}	0.0656 **	0.0627 **	17.9
Factor4*D8	k_{48}	0.0121	——	——
Factor5	k_{50}	-0.1383 **	-0.1119 **	749.8
Factor5*D1	k_{51}	0.0416 *	——	——
Factor5*D2	k_{52}	0.0732 **	0.0495 **	34.0
Factor5*D3	k_{53}	-0.0585 **	-0.0777 **	92.2
Factor5*D4	k_{54}	-0.0146	-0.0428 **	19.9
Factor5*D5	k_{55}	0.0328 **	——	——
Factor5*D6	k_{56}	0.0438 **	0.0166 *	4.2
Factor5*D7	k_{57}	-0.0228	-0.0291 *	4.6
Factor5*D8	k_{58}	-0.0450	-0.0500 *	6.1
Factor6	k_{60}	-0.0218 **	——	——
Factor6*D1	k_{61}	0.1118 **	0.1081 **	44.8
Factor6*D2	k_{62}	0.1481 **	0.1175 **	163.6
Factor6*D3	k_{63}	0.0484 **	0.0280 **	11.7
Factor6*D4	k_{64}	0.0250	——	——

Factor6*D5	k_{65}	0.0834 **	0.0631 **	89.7
Factor6*D6	k_{66}	0.0802 **	0.0546 **	29.5
Factor6*D7	k_{67}	0.0374	0.0274 *	5.4
Factor6*D8	k_{68}	0.0013	_____	_____
Factor7	k_{70}	0.0476 **	0.0490 **	308.4
Factor7*D1	k_{71}	0.0138	_____	_____
Factor7*D2	k_{72}	-0.0249 *	-0.0273 **	7.2
Factor7*D3	k_{73}	-0.0329 **	-0.0249 **	9.3
Factor7*D4	k_{74}	-0.0015	_____	_____
Factor7*D5	k_{75}	0.0020	_____	_____
Factor7*D6	k_{76}	-0.0239 *	-0.0255 **	7.3
Factor7*D7	k_{77}	-0.0172	_____	_____
Factor7*D8	k_{78}	-0.0275	_____	_____
Factor8	k_{80}	-0.0445 *	-0.0504 **	293.7
Factor8*D1	k_{81}	-0.0168	_____	_____
Factor8*D2	k_{82}	0.0159	0.0179 **	9.7
Factor8*D3	k_{83}	0.0001	_____	_____
Factor8*D4	k_{84}	-0.0901 *	-0.0978 **	9.6
Factor8*D5	k_{85}	-0.0105	_____	_____
Factor8*D6	k_{86}	0.1063 **	0.1124 **	11.5
Factor8*D7	k_{87}	-0.0844 *	-0.0868 **	16.3
Factor8*D8	k_{88}	0.0475	_____	_____
Factor9	k_{90}	-0.0739 **	-0.0665 **	103.4
Factor9*D1	k_{91}	0.0436	_____	_____
Factor9*D2	k_{92}	0.0827 **	0.0657 **	66.3
Factor9*D3	k_{93}	0.0168	_____	_____
Factor9*D4	k_{94}	-0.1106 **	-0.1290 **	26.4
Factor9*D5	k_{95}	0.0686 **	0.0573 **	56.2
Factor9*D6	k_{96}	0.0444 **	0.0330 **	10.4
Factor9*D7	k_{97}	-0.2003 **	-0.1243 **	12.7
Factor9*D8	k_{98}	-0.0804	_____	_____
Factor10	$k_{10,0}$	0.0002	_____	_____

Factor10*D1	$k_{10,1}$	0.0009	_____	_____
Factor10*D2	$k_{10,2}$	0.0303 **	0.0310 **	13.7
Factor10*D3	$k_{10,3}$	0.0145	0.0156 *	5.4
Factor10*D4	$k_{10,4}$	-0.0028	_____	_____
Factor10*D5	$k_{10,5}$	0.0185 **	0.0181 **	11.1
Factor10*D6	$k_{10,6}$	0.0145	_____	_____
Factor10*D7	$k_{10,7}$	0.0143	_____	_____
Factor10*D8	$k_{10,8}$	-0.0378	_____	_____
Factor11	$k_{11,0}$	-0.0115	_____	_____
Factor11*D1	$k_{11,1}$	-0.0148	_____	_____
Factor11*D2	$k_{11,2}$	0.0304 **	0.0195 **	15.0
Factor11*D3	$k_{11,3}$	0.0308 **	_____	_____
Factor11*D4	$k_{11,4}$	-0.0319 *	-0.0416 **	9.3
Factor11*D5	$k_{11,5}$	0.0191	_____	_____
Factor11*D6	$k_{11,6}$	-0.0092	-0.0237 *	4.4
Factor11*D7	$k_{11,7}$	-0.0301	_____	_____
Factor11*D8	$k_{11,8}$	-0.0524	_____	_____
Factor12	$k_{12,0}$	0.0251 **	0.0301 **	135.1
Factor12*D1	$k_{12,1}$	0.0073	_____	_____
Factor12*D2	$k_{12,2}$	-0.0291 **	-0.0344 **	18.3
Factor12*D3	$k_{12,3}$	-0.0029	_____	_____
Factor12*D4	$k_{12,4}$	-0.0059	_____	_____
Factor12*D5	$k_{12,5}$	0.0048	_____	_____
Factor12*D6	$k_{12,6}$	0.0087	_____	_____
Factor12*D7	$k_{12,7}$	0.0030	_____	_____
Factor12*D8	$k_{12,8}$	-0.0282	_____	_____

F Value	103.86	211.46	R^2 .8013	.7953
F Probability	.0001	.0001	Adjusted R^2	.7936

** Significant at the .01 level
* Significant at the .05 level

4. Regression Statistics for the 1980-1990 Model, Equation 5.6

Independent Variables/ Term		Parameter Estimate - Full Model	Parameter Estimate - Stepwise Model	F
Intercept	k_{00}	8.6568 **	8.6529 **	
YEAR	k_{01}	0.1107 **	0.1073 **	312.8
Factor1	k_{10}	-0.0725 **	-0.0722 **	584.5
Factor1*YEAR	k_{11}	-0.0195 **	-0.0189 **	21.2
Factor2	k_{20}	-0.0094 **	-0.0088 **	9.7
Factor2*YEAR	k_{21}	0.0169 **	0.0131 **	9.9
Factor3	k_{30}	0.1135 **	0.1148 **	1621.0
Factor3*YEAR	k_{31}	0.0436 **	0.0390 **	77.8
Factor4	k_{40}	-0.0434 **	-0.0465 **	551.7
Factor4*YEAR	k_{41}	-0.0065	_____	_____
Factor5	k_{50}	-0.0370 **	-0.0374 **	164.8
Factor5*YEAR	k_{51}	0.0134 **	0.0143 **	13.6
Factor6	k_{60}	-0.0918 **	0.0913 **	1818.6
Factor6*YEAR	k_{61}	-0.0001	_____	_____
Factor7	k_{70}	0.0805 **	0.0789 **	1634.8
Factor7*YEAR	k_{71}	-0.0028	_____	_____
Factor8	k_{80}	-0.0388 **	-0.0387 **	210.4
Factor8*YEAR	k_{81}	-0.0731 **	-0.0736 **	324.7
Factor9	k_{90}	0.0591 **	0.0550 **	779.2
Factor9*YEAR	k_{91}	-0.0083 *	_____	_____

F Value	677.81	886.64
F Probability	.0001	.0001
R^2	.6967	.6958

**   Significant at the .01 level
*   Significant at the .05 level

References

Agnew, J. 1987. *The United States in the World Economy: A regional Geography.* Cambridge University Press, Cambridge

Andrews, R.B. 1962. *Urban Growth and Development. A Problem Approach.* Simmons-Boardman Pub. Co., New York

Anglin, R. and B. Holcomb. 1992. "Poverty in Urban America: Policy Options." *Journal of Urban Affairs* 14 (3/4): 447-468.

Anon. 1989. "New Permanence of Poverty." *Society* 26 (Jan-Feb): 2-4.

Appalachian Regional Commission. 1966a. *Evaluation of Timber Development Organizations.* Appalachian Research Report 1, MacDonald Associates Inc., Washington D.C.

Appalachian Regional Commission. 1966b. *Industrial Location Research Studies: Summary and Recommendations.* Appalachian Regional Commission, Research Report 4.

Appalachian Regional Commission. 1972. *The Appalachian Experiment 1965-1970.* Appalachian Regional Commission.

Appalachian Regional Commission. 1977. *Appalachia - An Economic Report.* Appalachian Regional Commission, Washington D.C.

Appalachian Regional Commission. 1985. *Appalachia. Twenty Years of Progress.* Appalachian Regional Commission, Washington D.C.

Appelbaum, D.K. 1977. "The Level of the Poverty Line: A Historical Survey." *Social Service Review* 51 (September): 514-523.

Appelbaum, E. and P. Albin 1990 Shifts in Employment, Occupational Structure, and Educational Attainment. In *Skills, Wages, and Productivity in the service Sector.* T. Noyelle Ed., Westview Press, Boulder, Colorado: 31-66

Armstrong-Cummings, K. 1986. Strategies for a Stustainable Future in Appalachia. In *Proceedings from the 1986 Conference on Appalachia*. Appalachian Center for the Land and Economy of Appalachia. University of Kentucky, Lexington. Oct. 30th-31st: 9-15.

Arnold, V.L. 1980. *Alternatives to Confrontation. A National Policy Toward Regional Change*. Lexington Books, Lexington, MA.

Atkinson, A.B. 1987. "On the Measurement of Poverty." *Econometrica* 55 (4): 749-764

Auletta, K. 1982. *The Underclass*. Random House, New York.

Averitt, R.T. 1968. *The Dual Economy: The Dynamics of American Industry Structure*. Norton, New York.

Axinn, J. and Stern, M.J. 1988. *Dependency and Poverty*. Lexington Books, P.C. Heath & Co., Lexington, Massachusetts.

Barnet, R.J. and R.E. Muller. 1974. *Global Reach: The Power of Multinational Corporations*. Simon and Schuster, New York.

Bassett, J.E. 1973. *Regional Delineation of Poverty Levels in Appalachia*. University of Kentucky, Ph.D. Dissertation.

Batchelder, A.B. 1966. *The Economics of Poverty*. John Wiley and Sons, Inc., N.Y.

Beauregard, R.A. 1988. "In the Absence of Practice: The Locality Research Debate." *Antipode* 20 (1): 52-59.

Bednarzik, R 1990. A Special Focus on Employment Growth in Business Services and Retail Trade. In *Skills, Wages, and Productivity in the service Sector*. T. Noyelle Ed., Westview Press, Boulder, Colorado: 67-79

Beeghley, L. 1983. *Living Poorly in America*. Praeger, New York.

Belcher, J.C. 1962. Population Growth and Characteristics. In *The Southern Appalachian Region. A Survey*. Ed. T.R. Ford. University of Kentucky Press, Lexington, KY: 37-53.

Bethell, T.N. 1972. Conspiracy in Coal. In *Appalachia in the Sixties. Decade of Reawakening*. Eds. D. Walls and J.B. Stephenson. The University of Kentucky Press, Lexington, KY: 76-91.

Bethell, T.N., P. Gish, and T. Gish. 1972. Kennedy Hears of Need. In *Appalachia in the Sixties. Decade of Reawakening*. Eds. D. Walls and J.B. Stephenson. The University of Kentucky Press, Lexington, KY: 62-68.

Bishop, C.E. 1969. Rural Poverty in the Southeast. In *Rural Poverty and Regional Progress in an Urban Society*. Chamber of Commerce of the United States, Task Force on Economic Growth and Opportunity, Fourth Report: 75-92.

Block, F. 1990. *Postindustrial Possibilities*. University of California Press, Berkeley.

Bluestone, B. and B. Harrison. 1982. *The Deindustrialization of America*. Basic Books, New York.

Bonacich, E. 1976. "Advanced Capitalism and Black/White Race Relations in the United States: A split Labor Market Interpretation." *American Sociological Review* 41: 34-51.

Booth, C. 1970. *Labour and the Life of the People*. AMS Press, Inc., New York.

Bradshaw, M. 1992. *The Appalachian Regional Commission. Twenty-five Years of Government Policy*. University of Kentucky Press, Lexington

Bray, R. 1995. So How Did I Get Here? In *America's New War on Poverty*. Ed Robert Lavelle. KOED Books, San Francisco: 18-29

Bremner, R.H. 1992. *The Discovery of Poverty in the United States*. Transaction Publishers, New Brunswick, N.J.

Brooks, D.B. 1972. Strip Mining in Eastern Kentucky. In *Appalachia in the Sixties. Decade of Reawakening*. Eds. D. Walls and J.B. Stephenson. The University of Kentucky Press, Lexington, KY: 119-129.

Brunn, S.D. and Wheeler, J.O. 1971. "Spatial Dimensions of Poverty in the United States." *Geografiska Annaler* 53B (1): 6-15.

Bureau of National Affairs. 1988. *Consumer Price Index*. Bureau of National Affairs, Inc., Washington D.C., Policy and Practice Series.

Burton, C.E. 1992. *The Poverty Debate*. Greenwood Press, Westport, Connecticut.

Carnoy, M. 1984. *The State and Political Theory*. Princeton Universtity Press, Princeton, New Jersey.

Carson, C.B. 1991. *The War on the Poor*. American Textbook Committee, Birmingham, Alabama.

Casetti, E. 1972. "Generating Models by the Expansion Method: Applications to Geographical Research." *Geographical Analysis* 4 (1): 81-91.

Casetti, E. and J.P. Jones III 1983. "Regional Shifts in the
        Manufacturing Productivity Response to Output Growth:
        Sunbelt versus Snowbelt." *Urban Geography* 4: 285-301.

Caudill, H.M. 1972. Jaded Old Land of Bright New Promise. In
        Appalachia in the Sixties. Decade of Reawakening. Eds. D.
        Walls and J.B. Stephenson. The University of Kentucky Press,
        Lexington, KY: 240-246.

Center on Budget and Policy Priorities. 1988a. *National Overview*. 236
        Massachusetts Ave, Washington D.C. (April).

Center on Budget and Policy Priorities. 1988b. *Holes in the Safety
        Nets*. Poverty Programs and Policies in the States. 236
        Massachusetts Ave, Washington D.C. (Spring).

Checchi and Company. 1969. *Capital Resources In the Central
        Appalachian Region*. Appendix C to 'A Strategy for
        Development in Central Appalachia'. A Report to the
        Appalachian Regional Commission, Washington D.C.

Chinitz, B. 1969. The Regional Aspects of Poverty. In *Rural Poverty
        and Regional Progress in an Urban Society*. Chamber of
        Commerce of the United States, Task Force on Economic
        Growth and Opportunity, Fourth Report: 93-104.

Cochrane, A. 1987. "What a Difference Place Makes: The New
        Structuralism of Locality." *Antipode* 19 (3): 354-363.

Cohen, I.J. 1987. Structuration Theory and Social Praxis. In *'Social
        Theory Today'*, A. Giddens and J.H. Turner Eds, Stanford
        University Press, Stanford, California: 273-308.

Cohen, I.J. 1989. *Structuration Theory*. MacMillan, Basingstoke,
        Hampshire.

Coleman, K. 1991. *A History of Georgia*. University of Georgia Press,
        Athens.

Conference on Economic Progress. 1962. *Poverty and Deprivation in
        the U.S. The Plight of Two-fifths of a Nation*. Conference on
        Economic Progress, 1001 Connecticut Ave. N.W.,
        Washington D.C.

Cooke, P. 1987a. "Clinical Inference and Geographic Theory."
        Antipode 19 (1): 69-78.

Cooke, P. 1987b. "Individuals, Localities, and Postmodernism."
        Environment and Planning D: Society and Space 5 (4): 408-
        412.

Cooke, P. 1989. "The Contested Terrain of Locality Studies."*Tijdschift voor Economische en Sociale Geografie* 80 (1): 14-29.

Coraggio, J.L. 1983 Social Spaceness and the Concept of Region. In *Regional Analysis and the New International Division of Labor.* F. Moulaert and P. Wilson Salinas, eds., Kluwer Nijhoff Pub., Boston.

Cowell, F.A. 1977. *Measuring Inequality. Techniques for the Social Sciences.* Halsted Press - John Wiley and Sons, Inc., N.Y.

Cox, K.R. 1989. The Politics of Turf and the Question of Class. In *The Power of Geography. How Territory Shapes Social Life.* Eds. J. Wolch and M. Dear. Unwin Hyman, Boston, MA: 61-90.

Cox, K.R. 1993. "The Local and the Global in the New Urban Politics: A Critical View." *Environment and Planning D: Society and Space* 11: 433-448.

Craib, I. 1984. *Modern Social Theory.* St Martin's Press, New York.

Day, G. 1994. The Reconstruction of Wales and Appalachia: Development and Regional Identity. In *Appalachia in an International Context.* Eds, Phillip J. Obermiller and William W. Philliber. Praeger, Westport CT: pp. 45-65

Delhausse, B., A. Luttgens, and S. Perelman. 1993 "Comparing Measures of Poverty and Relative Deprivation. An Example of Belgium." *Journal of Population Economics* 6: 83-102.

Davies, W.K.D. and D.T. Herbert 1993. *Communities within Cities: An Urban Social Geography.* Belhaven Press, London.

Dear, M.J. and J.R. Wolch. 1989. How Territory Shapes Social Life. In *The Power of Geography. How Territory Shapes Social Life.* Eds. J. Wolch and M. Dear. Unwin Hyman, Boston, MA: 3-18.

Deavers, K.L. 1980. Rural Conditions and Regional Differences. In *Alternatives to Confrontation. A National Policy Toward Regional Change.* Ed V. L. Arnold. Lexington Books, Lexington, MA.

Devine, J.A. M. Plunkett, and J.D. Wright. 1992. "The Chronicity of Poverty: Evidence from the PSID, 1968-1987." *Social Forces* 70 (3): 787-812.

Dhillon, J.S. and M.R. Howie. 1986. *Dimensions of Poverty in the Rural South.* Florida Agricultural and Mechanical University, Tallahassee.

DiLulio, J.J. 1989. "The Impact of Inner-city Crime." *The Public
    Interest* 96: 28-46

Division of Community Services. 1981. *The Winning Side of the War
    on Poverty. 1964-1980.* Department of Human Services, Little
    Rock, (August).

Dobelstein, A.A. 1987. *Examining Poverty.* North Carolina Poverty
    Project, Greensboro.

Dudenhefer, P. 1993. "Poverty in the Rural United States." *Focus*
    (University of Wisconsin-Madison Institute for Research on
    Poverty) 15 (1): 37-45.

Dudley, W. 1988. *Poverty. Opposing Viewpoints.* Greenhaven Press,
    St Paul, Minnesota.

Duncan, C.M. 1986. Myths and Realities of Appalachian Poverty:
    Public Policy for Good People Surrounded by a Bad Economy
    and Bad Politics. In *Proceedings from the 1986 Conference on
    Appalachia.* Appalachian Center for the Land and Economy of
    Appalachia. University of Kentucky, Lexington. Oct. 30th-
    31st: 25-32.

Duncan, O.D. 1969. Inheritance of Poverty or Inheritance of Race. In
    On Understanding Poverty. Ed. D.P. Moynihan, Basic Books,
    Inc., New York: 85-110.

Eberstadt, N. 1988. "Economic and Material Poverty in the U.S."
    Public Interest 90 (Winter): 50-65.

Eggers, M.L. and D.S. Massey. 1991. "The Structural Determinants of
    Urban Poverty: A Comparison of Whites, Blacks, and
    Hispanics." *Social Science Research* 20 (3): 217-255

Eggers, M.L. and D.S. Massey. 1992. "A Longitudinal Analysis of
    Urban Poverty: Blacks in U.S. Metropolitan Areas Between
    1970 and 1980." *Social Science Research* 21: 175-203.

Elkin, S.L. 1987 *City and Regime in the American Republic.*
    University of Chicago Press.

Ferman, L.A., J.L. Kornbluh, and A. Haber. Editors. 1969.   *Poverty
    in America. A Book of Readings.* University of Michigan
    Press.

Evans, P. 1986. Transnational Linkages and the Economic Role of the
    State. In *Bringing the State Back In.,* Eds P. Evans, D.
    Rueschemeyer, and T. Skocpol. Cambridge University Press,
    Cambridge: 192-226.

Fitchen, J.M. 1981. *Poverty in Rural America: A Case Study.*
Westview Special Studies in Contemporary Social Issues,
Boulder, Colorado.

Ford, A.M. 1973. *Political Economics of Rural Poverty in the South.*
Ballinger Publishing Co., Cambridge, MA.

Ford, T.R. 1964. *Health and Demography in Kentucky.* University of
Kentucky Press, Lexington, KY.

Ford, T.R. 1969. Rural Poverty in the United States. In *Rural Poverty
and Regional Progress in an Urban Society.* Chamber of
Commerce of the United States, Task Force on Economic
Growth and Opportunity, Fourth Report.

Fosler, R.S. and R.A. Berger 1982. *Public-Private Partnership in
American Cities.* Lexington Books, Mass.

Foster, S.A. 1991. "The Expansion Method: Implications for
Geographic Research." *Professional Geographer* 43 (2): 131-
142.

Frank, A.G. 1979. *The Development of Underdevelopment.* Monthly
Review Press, New York.

Frank, A.G. 1987. Global Crisis and Transformation. In *International
Capitalism and Industrial Restructuring: A Critical Analysis.*
Ed. R. Peet. Allen and Unwin, Boston, MA: 293-312.

Frey, W.H. 1993. *Race, Class and Poverty Polarization Across Metro
Areas and States: Population Shifts and Migration Dynamics.*
Population Studies Center, University of Michigan Research
Report No.93-293, Sept.

Friedland, R. 1983. "The Politics of Profit and the Geography of
Growth." *Urban Affairs Quarterly* 19 (1): 41-54.

Frobel, F., J. Heinrichs, and O. Kreye. 1977. "The Tendency Towards a
New International Division of Labor." *Review* 1 (1): 73-88.

Fuchs, V.R. 1992. "Poverty and Health: Asking the Right Questions. "
The American Economist 36 (2): 12-18.

Gallaway, L and Vedder, R. 1985. *'Suffer the Little Children': The True
Casualties of the War on Poverty.* Joint Economic
Committee, U.S. Congress. War on Poverty: Victory or
Defeat. Hearing before the Subcommittee on Monetary and
Fiscal Policy, June 20th.

Gallaway, L., Vedder, R. and Foster, T. 1985. *The 'New' Structural Poverty: A Quantitative Analysis*. Joint Economic Committee, U.S. Congress. War on Poverty: Victory or Defeat. Hearing before the Subcommittee on Monetary and Fiscal Policy, June 20th.

Gaventa, J. 1994 From the Mountains to the Maquiladoras. A Case Study of Capital Flight and its Impact on Workers. In *Appalachia in an International Context*. Eds, Phillip J. Obermiller and William W. Philliber. Praeger, Westport CT: pp. 165-176

Gaventa, J. 1980. *Power and Powerlessness: Quiescence and Rebellion in an Appalachian Valley*. Clarendon Press, Oxford.

Giarratani, F. and C. Rogers. 1991. "Some Spatial Aspects of Poverty in the USA." *Population Research and Policy Review* 10: 213-234.

Giddens, A. 1984. *The Constitution of Society: Outline of the Theory of Structuration*. Polity Press/University of California Press, Berkeley.

Giddens, A. and J.H. Turner. 1987. *Social Theory Today*. Stanford University Press, Stanford, California.

Gilbert, A. 1988 "The New Regional Geography in English and French-speaking Countries." *Progress in Human Geography* 12: 208- 28

Ginsberg,E., T.J. Noyelle and T.M. Stanback. 1986. *Technology and Employment. Concepts and Clarifications*. Westview Press, Boulder, COlorado

Goldstein, R. 1986. *Economic Recovery Fails to Reduce Poverty Rates to Pre-recession Levels. Gap Widens Further Between Rich and Poor*. Center on Budget and Policy Priorities, Washington D.C.

Goldstein, R. and S.M. Sachs. 1983. *Applied Poverty Research*. Rowman and Allanheld, New Jersey.

Good, P. 1972. Coal Beds of Sedition. In *Appalachia in the Sixties. Decade of Reawakening*. Eds. D. Walls and J.B. Stephenson. The University of Kentucky Press, Lexington, KY: 184-193

Gordon, D. 1988. "The Global Economy: New Edifice or Crumbling Foundations." *New Left Review*

Gordon, D., R. Edwards, and M. Reich. 1982. The Historical Transformation of Labor: An Overview. In *Segmented Work, Divided Workers*, Cambridge University Press, Cambridge: 1-17.

Gottschalk, P. 1985. *The Successes and Limitations of the War on Poverty and the Great Society Programs*. Joint Economic Committee, U.S. Congress. Hearing before the Subcommittee on Monetary and Fiscal Policy 'War on Poverty: Victory or Defeat' June 20th.

Granovetter, M. 1985. "Economic Action and Social Structure." *American Journal of Sociology* 91: 481-510.

Granovetter, M. 1986. Labor Mobility, Internal Markets, and Job Matching: A Comparison of the Sociological and Economic Approaches. In *Research in Social Stratification and Mobility Vol.5*. R.V.Robinson, Editor. JAI Press.

Gray, R. and J.M. Peterson. 1974. *Economic Development of the United States*. Richard D. Irwin, Inc., Homewood, Illinois.

Greene, R. 1991. "Poverty Concentration Measures and the Urban Underclass." *Economic Geography* 67 (3): 240-252.

Greenstein, R. 1985. *Prepared Statement*. Joint Economic Committee, U.S. Congress. Hearing before the Subcommittee on Monetary and Fiscal Policy 'War on Poverty: Victory or Defeat' June 20th

Greenstein, R. 1986. *Numbers and Need: The Outlook for Working Families and the Poor*. Agenda '87: Revolutionalizing State Economic Policy, Midwest Regional Conference, Chicago, November 21st-22nd.

Gregson, N. 1987. "The CURS Initiative: Some Further Comments." *Antipode* 19 (3): 364-370.

Gugliotta, G. 1994 "The Persistence of Poverty." *Washington Post* 11 (10) Jan 3rd-9th: 6-9

Guinness, P. and M. Bradshaw. 1985. *North America: A Human Geography*. Barnes and Noble, Totowa, N.J.

Hamilton, C.H. 1962. Health and Health Services. In *The Southern Appalachian Region. A Survey*. Ed. T.R. Ford. University of Kentucky Press, Lexington, KY: 219-244.

Hamm, K. 1973. *Appalachia - An Economic Report*. Trends in Employment and Income. Appalachian Regional Commission, Supplement.

Hanson, S. and G. Pratt. 1991. "Job Search and Occupational Segregation of Women." *Annals of the Association of American Geographers* 81 (2): 229-253.

Harrington, M. 1963. *The Other America: Poverty in the United States.* Penguin Books, Baltimore

Harrington, M. 1984. *The New American Poverty.* Holt, Rinehart and Winston, N.Y.

Harrington, M. 1987. Causes and Effects of Poverty and Possible Solutions. In *Poverty and Social Justice: Critical Perspectives* Ed F. Jimenez. Bilingual Press, Tempe, Arizona: 27-38.

Harrington, M. 1988. "The First Steps - and a Few Beyond." *Dissent*: 44-47 & 55, Winter.

Hartman, H. 1976. Capitalism, Patriarchy, and Job Segregation by Sex. In *Women and the Workplace: The Implications of Occupational Segregation.* M. Blaxall and B. Reagan, Editors., University of Chicago Press, Chicago: 137-169.

Harvey, D. 1987. "Three Myths in Search of a Reality in Urban Studies." *Environment and Planning D: Society and Space* 5 (4): 367-376.

Harvey, D. 1985a. *Consciousness and the Urban Experience.* Johns Hopkins University Press, Baltimore, Maryland.

Harvey, D. 1985b. The Geopolitics of Capitalism. In *Social Relations and Spatial Structures.* Eds. D. Gregory and J. Urry,. St. Martin's Press, New York: 128-163.

Harvey, D. 1972. *Social Justice and Spatial Systems.* Antipode Monographs in Social Geography. Geographical Perspectives on American Poverty: 87-106.

Haveman, R.H. 1987. *Poverty Policy and Poverty Research. The Great Society and the Social Sciences.* University of Wisconsin Press, Madison.

Haveman, R. 1992/1993. "Changing the Poverty Measure: Pitfalls and Potential Gains." *Focus* (University of Wisconsin-Madison Institute for Research on Poverty) 14 (3): 24-29.

Hepburn, L.R. 1992. *Contemporary Georgia.* Institute of Government, Athens, Georgia.

Higgins, J. 1978. *The Poverty Business. Britain and America.* Basil Blackwell, Oxford.

Hirschl, T.A. and M.R. Rank. 1991 "The Effect of Population Density on Welfare Participation." *Social Forces* 70 (1): 225-235.

Hodson, R.D. 1984 "Companies, Industries, and the Measurement of Economic Segmentation." *American Sociological Review* 49: 335-348

Hodson, R., P.G. Schervish, and R.Stryker. 1988. "Class Interests and Class Fractions in an Era of Economic Decline." *Research in Politics and Society* 3: 191-220.

Hoffman, E.P. 1991. "Aid to Families with Dependent Children and Female Poverty." *Growth and Change* 22 (2): 36-47.

Holman, R. 1978. *Poverty. Explanations of Social Deprivation*. St. Martin's Press, New York.

Hoppe, R.A. 1991. "Defining and Measuring Poverty in the Nonmetropolitan United States Using the Survey of Income and Program Participation." *Social Indicators Research*, 24: 123-151.

Hormats, R.D. 1984. Unemployment and Growth in the Western Economies. In *Unemployment and Growth in the Western Economies*, A.J. Pierre Ed., New York University Press: 1-13

Hughes, J.T. 1991. "Evaluation of Local Economic Development: A Challenge for Policy Research." *Urban Studies* 28 (6): 909-918

Hughes,M.A. 1990. "Formation of the Impacted Ghetto: Evidence from Large Metropolitan Areas." *Urban Geography* 11 (3): 265-284.

Jackson, G., G. Masnick, R. Bolton, S. Bartlett, and J. Pitkin. 1983. *Regional Diversity. Growth in the United States, 1960-1990*. Auburn House Publishing Company, Boston, Massachusetts.

Johnson,H.J. 1991. *Dispelling the Myth of Globalization. The Case for Regionalization*. Praeger, New York.

Johnston, R.J. 1990. Economic and Social Policy Implementation and Outputs. In *Dimensions of United States Social Policy*, J.E. Kodras and J.P. Jones Eds., Edward Arnold, London.

Jones, J.P. 1984. "A Spatially-Varying Parameter model of AFDC Participation: Empirical Analysis Using the Expansion Method." *Professional Geographer* 36 (4): 455-461.

Jones, J.P. 1987. "Work, Welfare, and Poverty Among Black Female Headed Families." *Economic Geography* 63: 20-34.

Jones, J.P. and J.E. Kodras. 1990. "Restructured Regions and Families: The Feminization of Poverty in the United States." *Annals of the Association of American Geographers*, 80 (2): 163-183.

Jordan, B., S. James, H. Kay, and M. Redley. 1992. *Trapped in Poverty? Labour-market decisions in Low Income Households.* Routledge, London.

Kahan, D. and K. McKeown. 1986. *New Census Data Shows Widening Gap Between Wealthiest and Other Americans.* Center on Budget and Policy Priorities, Washington D.C., (August).

Kaplan, C.P. and T.L. Van Valey 1980. *Census 80: Continuing the Factfinder Tradition.* U.S. Department of Commerce, Bureau of the Census, Jan.

Kasarda, J.D. 1989. "Urban Industrial Transition and the Underclass." *The Annals of the American Academy of Political and Social Science* 501: 26-47.

Kasarda, J.D. 1990. "Structural Factors Affecting the Location and Timing of Urban Underclass Growth." *Urban Geography* 11 (3): 234-264.

Katz, M.B. 1983. *Poverty and Policy in American History.* Academic Press, N.Y.

Katz, M.B. 1989. *The Undeserving Poor.* Pantheon Books, New York.

Katz, M.B. 1993. *The "Underclass" Debate.* Princeton University Press, New Jersey.

Kirschenman, J. and K.M. Neckerman. 1991. "We'd Have to Hire Them But..." The Meaning of Race for Employers. In *The Urban Underclass.* Eds C. Jencks and P.E. Peterson. The Brookings Institute, Washington D.C.

Knox, P.L. 1974. "Spatial Variations in the Level of Living in England and Wales in 1961." *Transactions of the Institute of British Geographers,* 62: 1-23.

Knox, P.L. 1990. "The New Poor and a New Urban Geography." *Urban Geography* 11 (3): 213-216.

Knox, P.L. 1991. "The Restless Urban Landscape: Economic and Sociocultural Change and the Transformation of Washington,DC." *Annals of The Association of American Geographers* 81 (2): 181-209.

Kodras, J.E. 1992. *Breadlines. In Geographical Snapshots of North America,* Ed D.G. Janelle, The Guilford Press, N.Y.: 103-107.

Kodras, J.E. and J.P. Jones III. 1990. *Dimensions of United States Social Policy.* Edward Arnold, London.

Kublawi, S. 1986. The Economy of Appalachia in the National
   Context. In *Proceedings from the 1986 Conference on
   Appalachia*. Appalachian Center for the Land and Economy of
   Appalachia. University of Kentucky, Lexington. Oct. 30th-
   31st: 16-24.

Larner, J. and I. Howe. 1970. *Poverty: Views from the Left*. William
   Morrow and Co., Inc., N.Y.

Laws, G. 1989. Privatization and Dependency on the Local Welfare
   State. In T*he Power of Geography. How Territory Shapes
   Social Life*. Eds. J. Wolch and M. Dear. Unwin Hyman,
   Boston, MA: 238-257.

Lawson, V. and L. Staeheli. 1990. "Realism and the Practice of
   Geography." *The Professional Geographer* 42 (1): 13-20.

Leacock, E.B. 1971. *The Culture of Poverty. A Critique*. Simon and
   Schuster, New York.

League of Women Voters. 1988. *Unmet Needs: The Growing Crisis in
   America*. Voters Education Fund Publication 853.

Lefebvre, H. 1991. *The Production of Space*. Basil Blackwell, Oxford.

Levitan, S.A. and I. Shapiro. 1987. *Working But Poor*. Johns Hopkins
   University Press.

Lewis, O. 1969. The Culture of Poverty. In *On Understanding Poverty*.
   Ed. D.P. Moynihan, Basic Books, Inc., New York: 187-200.

Linder, P.H. and C.P. Kindleberger. 1982. *International Economics*.
   7th Edition. R.D. Irwin, Homewood, Illinois.

Lipietz, A. 1986. New Tendencies in the International Division of
   Labor: Regimes of Accumulation and Modes of Regulation. In
   *Production, Work, Territory: The Geographical Anatomy of
   Industrial Capitalism*. A. Scott and M. Stroper Editors, Allen
   and Unwin, Boston: 16-40.

Logan, J.R. and H.L. Molotch 1987. *Urban Fortunes. The Political
   Economy of Place*. University of California Press, Berkeley

Lovering, J. 1989. "Postmodernism, Marxism, and Locality Research:
   The Contribution of Critical Realism to be Debate." *Antipode*
   21 (1) 1-12.

Lund, L. 1976. *Business-Government Partnership in Local Economic
   Development Planning*. The Conference Board Inc., New
   York.

MacDonald, D. 1969. The Invisible Poor. In *Poverty in America. A
   Book of Readings*, University of Michigan Press.

MacLachlan, G. 1974. *The Other Twenty Percent. A Statistical Analysis of Poverty in the South.* Southern Regional Council, Inc., 52 Fairlie St. N.W., Atlanta Ga 30303.

Macnicol, J. 1987. "In Pursuit of the Underclass." *The Journal of Social Policy* 16 (3): 293-318.

Mangold, W.D., W.A. Schwab, and D.E. Ferritor. 1980. *Newborn Mortality in Arkansas.* University of Arkansas, Department of Sociology, Final Report to the Arkansas Regional Prenatal Program, June.

Manski, C.F. 1992/1993. "Income and Higher Education." *Focus* (University of Wisconsin-Madison Institute for Research on Poverty) 14 (3): 14-19

Marx, K. 1977. *Capital, Volume One.* Vintage Books, New York.

Massey, D. 1984. *Spatial Divisions of Labour. Social Structures and the Geography of Production.* MacMillan, London.

Mattera, P. 1990. Prosperity Lost. Addison-Wesley Publishing Co., Inc., Reading MA.

McCombie, J.S.L. 1988a. "A Synoptic View of Regional Growth and Unemployment: I. The Neoclassical Theory." *Urban Studies* 25: 267-281

McCombie, J.S.L. 1988b. "A Synoptic View of Regional Growth and Unemployment: II. The Post-Keynesian Theory." *Urban Studies* 25: 399-417

McCormick, J. 1988. "America's Third World." *Newsweek* (Aug 8th): 20-24.

McLanahan, S. and I. Garfinkel. 1989. "Single Mothers, the Underclass and Social Policy." *The Annals of the American Academy of Political and Social Science* 501: 92-104.

McLernon, P.A. 1989. *Spatial Analysis of Poverty: A Case Study of the State of Georgia.* University of Georgia, Athens, M.A. Thesis.

McNutt, J.G. 1986. An Alternative Development Strategy for Appalachia's Future: Application of "Another Development" and "Sustainable Society" Themes to a Region in Crisis. In *Proceedings from the 1986 Conference on Appalachia.* Appalachian Center for the Land and Economy of Appalachia. University of Kentucky, Lexington. Oct. 30th-31st: 51-58.

Mead, L.M. 1989. "The Logic of Workfare: The Underclass and Work Policy." *The Annals of the American Academy of Political and Social Science* 501: 156-169.

Mead, L.M. 1992. *The New Politics of Poverty*. Basic Books, Inc., New York.

Mertz, P.E. 1978. *New Deal Policy and Southern Rural Poverty*. Louisiana State University Press, Baton Rouge.

Miliband, R. 1990. *Divided Societies*. Oxford University Press, Oxford.

Millstone, J.C. 1972. East Kentucky Coal Makes Profits for Owners, Not Region. In *Appalachia in the Sixties. Decade of Reawakening*. Eds. D. Walls and J.B. Stephenson. The University of Kentucky Press, Lexington, KY: 69-76.

Mingione, E. 1991. *Fragmented Societies. A Sociology of Economic Life Beyond the Market Paradigm*. Basil Blackwell, Oxford.

Moore, T. 1994 "Core-periphery Models, Regional Planning Theory, and Appalachian Development." *The Professional Geographer* 46 (3) : 316-331

Moos, A. and M. Dear. 1986. "Structuration Theory in Urban Analysis: 1. Theoretical Exegesis." *Environment and Planning A* 18: 231-251.

Morgan, J.N., M.H. David, W.J. Cohen, and H.E. Brazer. 1962. *Income and Welfare in the United States*. McGraw Hill Book Co., Inc.,N.Y.

Morrill, R.L. and E.H. Wohlenberg. 1971. *The Geography of Poverty in the United States*. McGraw-Hill Book Co., Problems Series.

Moynihan, D.P. 1989. "Towards a Post-industrial Social Policy." *The Public Interest* 96: 16-27.

Murray, C. 1984. *Losing Ground: American Social Policy 1950-1980*. Basic Books, New York

Murray, C. 1988a. "Poverty and Welfare: The Working Poor." *Current*: 8-15 (May).

Murray, C. 1988b. "What's So Bad About Being Poor?" *National Review*: 36-39 & 59, October 28.

Murray, C. 1992. "Causes, Root Causes, and Cures." *National Review*: 30-32, June 8.

Nathan, R.P. 1989. "Institutional Change and the Challenge of the Underclass." *The Annals of the American Academy of Political and Social Science* 501: 170-181.

Netzer, D. 1974. *Economics and Urban Problems*. Basic Books Inc., New York.

Newman, M. 1972. *The Political Economy of Appalachia. A Case Study in Regional Integration*. Lexington Books, Lexington, MA.

Nord, S. and R.G. Sheets. 1992. Service Industries and the Working Poor in Major Metropolitan Areas in the United States. In *Sources of Metropolitan Growth*. E.S. Mills and J.F. McDonald, Eds., Center for Urban Policy Research, New Brunswick, N.J.

Noyelle,T.J. 1987. *Beyond Industrial Dualism. Market and Job Segmentation in the New Economy*. Westview Press, Boulder Colorado.

Noyelle, T.J. and T.M. Stanback 1984. *The Economic Transformation of American Cities*. Rowman and Allanheld, Totowa, N.J.

Noyelle, T.J. and T.M. Stanback 1990. Productivity in Services: A Valid Measure of Economic Performance? In *Skills, Wages, and Productivity in the service Sector*. T. Noyelle Ed., Westview Press, Boulder, Colorado: 187-211

Offord, J. 1983. *State Welfare Expenditures and the Geography of Social Well-being in the US State of Georgia*. Queen Mary College, University of London, Occasional   Paper No.21.

O'Hare, W.P. 1985. "Poverty in America: Trends and New Patterns." *Population Bulletin* 40 (June): 2-43.

O'Hare, W.P. 1987. *America's Welfare Population: Who Gets What?* Population Reference Bureau, Population Trends and Public Policy, Occasional Paper, 13 September.

O'Hare, W.P. 1988. *The Rise of Poverty in Rural America*. Population Reference Bureau, Population Trends and Public Policy, Occasional paper 15, July.

O'Hare, W.P. 1989. "The Rise of Rural Poverty." *The Futurist* 31 (January-February): 45.

Ohlin, B.G. 1933. *Interregional and International Trade*. Harvard University Press, Cambridge, Massacusetts.

Oppenheimer, M. 1985. "White Collar Politics." *Monthly Review Press*, New York.

O'Regan, K. and M. Wiseman. 1990. "Using Birth Weights to Chart the Spatial Distribution of Urban Poverty." *Urban Geography* 11 (3): 217-233.

Ornati, O. 1969. Poverty in America. In *Poverty in America. A Book of Readings*, University of Michigan Press.

Orshansky, M. 1969. Counting the Poor: Another Look at the Poverty Profile. In *Poverty in America. A Book of Readings*, University of Michigan Press.

Osborne, G.L., M.V. Carter, and A.C. Butler. 1986. Appalachian Trends and the Liberal Arts College: A Perspective and Case Study Based on Power Theory. In *Proceedings from the 1986 Conference on Appalachia*. Appalachian Center for the Land and Economy of Appalachia. University of Kentucky, Lexington. Oct. 30th-31st: 59-69.

Oster, S.M., E.E. Lake, and C.G. Oksman. 1978. *The Definition and Measurement of Poverty. Vol.1*. A Review. Westview Special Studies in Applied Social Research, Westview Press, Boulder, Colorado.

Oyen, E. 1992. "Some Basic Issues in Comparative Poverty Research." *International Social Science Journal* 134: 615-626.

Paasi, A. 1986. "The Institutionalization of Regions: A Theoretical Framework for Understanding the Emergence of Regions and the Constitution of Regional Identity." *Fennia* 164 (1): 106-146.

Pandit, K. and E. Casetti 1989. "The Shifting Patterns of Sectoral Labor Allocation During Development: Developed versus Developing Countries." *Annals of the Association of American Geographers* 79: 329-344.

Pearson, N.K. 1994 Local Development Activities in Newfoundland and Central Appalachia. In *Appalachia in an International Context*. Eds, Phillip J. Obermiller and William W. Philliber. Praeger, Westport CT: pp. 91-110

Peet, R. 1972. "Some Issues in the Social Geography of American Poverty." *Antipode Monographs in Social Geography* 1, Geographical Perspectives on American Poverty: 1-16.

Peet, R. 1987. Industrial Restructuring and the Crisis of International Capitalism. In *International Capitalism and Industrial Restructuring: A Critical Analysis*. Ed. R. Peet. Allen and Unwin, Boston, MA: 9-32.

Peterson, J.M. 1962. "National Trends and Arkansas's Economic Development." *The Arkansas Economist* 4 (Spring): 6.

Peterson, P.E. 1981. *City Limits*. University of Chicago Press, Chicago,IL.

Piore, M. and C. Sable. 1984. *The Second Industrial Divide*. Basic Books, New York.

Piven, F.F. and R.A. Cloward. 1974. Regulating the Poor. In *The Poverty Establishment*. Ed. P. Roby. Prentice-Hall, Inc., Englewood Cliffs, New Jersey: 25-42.

Piven, F.F. and R.A. Cloward. 1982. *The New Class War: Reagan's Attack on the Welfare State and Its Consequences*. Pantheon Books, New York.

Plotnick, R.D. 1990. "The Impact of Transfers on Poverty." *Focus* 12 (3): 8-9.

Plotnick, R.D. and F. Skidmore. 1975. *Progress Against Poverty. A Review of the 1964 - 1974 Decade*. Academic Press, N.Y.

Pollard, K. 1992. "Income Down, Poverty Up in Latest CPS Reports." *Population Today* 20 (11): 5 & 9.

Porter, K. 1989. *Poverty in Rural America. A National Overview*. Center on Budget and Policy Priories, Washington D.C.

Pred, A. 1985. The Social Becomes the Spatial, the Spatial Becomes the Social: Enclosures, Social Change and the Becoming of Places in Skane. In *Social Relations and Spatial Structures*, Ed D.Gregory and D. Urry. St Martin's Press, N.Y.,: 337-365.

Pred, A. 1986. *Place, Practice and Structure*. Barnes and Noble Totowa, N.J.

Pudup, M.B. 1987. *Land Before Coal: Class and Regional Development in Southeast Kentucky*. Ph.D. Dissertation, University of California, Berkeley.

Quibria, M.G. 1991. "Understanding Poverty: An Introduction to Conceptual and Measurement Issues." *Asian Development Review* 9 (2): 90-112.

Radomski, A.L. and A.U. Mills. 1964. *Family Income and Related Characteristics Among Low-income Counties and States*. U.S. Department of Health, Education, and Welfare, Welfare Research Report 1 (September).

Reitsma, H.A. and J.M.G. Kleinpenning. 1989. *The Third World In Perspective*. Van Gorcum, Assen/Maastricht, Netherlands.

Renwick, T.J. and B.R. Bergmann. 1993. "A Budget Based Definition of Poverty with an Application to Single Parent Families." *The Journal of Human Resources* 28 (1): 1-24.

Reul, M.R. 1974. *Territorial Boundaries of Rural Poverty*. Center of Rural Manpower and Public Affairs & The Cooperative Extension Service, Michigan State University, East Lansing.

Rickets, E.R. and I.V. Sawhill. 1988. "Defining and Measuring the Underclass." *The Journal of Policy Analysis and Management* 7 (2): 316-325.

Roby, P. 1974. *The Poverty Establishment*. Prentice-Hall, Inc., Englewood Cliffs, New Jersey.

Rodgers, J.R. 1991. "Does the Choice of Poverty Index Matter in Practice." *Social Indicators Research* 24: 233-252.

Rodgers, J.R. and J.L. Rodgers. 1993. "Chronic Poverty in the United States." *The Journal of Human Resources* 28 (1): 25-54.

Ropers, R.H. 1991. *Persistent Poverty. The American Dream Turned Nightmare*. Insight Books, Plenum Press, N.Y.

Ross, C., S. Danziger, and E. Smolensky. 1987. "The Level and Trend of Poverty in the United States." *Demography* 24 (November): 587-99.

Rossi, P.H. and Z.D. Blum. 1969. Class, Status, and Poverty. In *On Understanding Poverty*. Ed. D.P. Moynihan, Basic Books, Inc., New York: 36-63.

Rossi, P.H. and J.D. Wright. 1989. "The Urban Homeless: A Portrait of Urban Dislocation." *The Annals of the American Academy of Political and Social Science* 501: 132-142.

Rowntree, B.S. and G.R. Lavers. 1951. *Poverty and the Welfare State*. Longmans, Green and Co., London.

Rubenstein, J.M. 1989. *The Cultural Landscape. An Introduction to Human Geography*. Macmillan Publishing Co., New York.

Rubinstein, E. 1989. "Losing More Ground." *National Review* 41 (May):13.

Ruggles, P. 1990. *Drawing the Line*. Urban Instutute Press, Washington D.C.

Ruggles, P. 1992. "Measuring Poverty." University of Wisconsin-Madison, Institute for Research on Poverty, *Focus* Vol 14, No.1 (Spring): 1-9.

Ryan, W. 1974. Blaming the Victim: Ideology Serves the Establishment. In *The Poverty Establishment*. Ed. P. Roby. Prentice-Hall, Inc., Englewood Cliffs, New Jersey: 171-179.

Sanderson, S. 1985. *The America in the New International Division of Labor*. Holmes and Meier, New York.

Sarre, P. 1987. "Realism in Practice." *Area* 19 (1): 3-10.

Sassen, S. 1988. *The Mobility of Labor and Capital: A Study in International Investment and Labor Flow*. Cambridge University Press, New York.

Sawhill, I.V. 1989. "The Underclass. An Overview." *Public Interest* 96: 3-15.

Sayer, A. 1984. *Method in Social Science. A Realist Approach.* Hutchinson, London.

Sayer, A. 1987. "Hard Work and its Alternatives." *Environment and Planning D: Society and Space* 5 (4): 395-399.

Scalf, H.P. 1972. *Kentucky's Last Frontier.* Pikeville College Press/Appalachian Studies Center, Pikeville, KY.

Scheller, M. 1995. On the Meaning of Plumbing and Poverty. In *America's New War on Poverty.* Ed Robert Lavelle. KOED Books, San Francisco: 118-122

Schoenberger, E. 1988. "From Fordism to Flexible Accumulation: Technology, Competitive Strategies, and International Location." *Environment and Planning D* 6: 245-262.

Schrag, p. 1972. The School and Politics. In *Appalachia in the Sixties. Decade of Reawakening.* Eds. D. Walls and J.B. Stephenson. The University of Kentucky Press, Lexington, KY: 219-224.

Schwarz, J.E. and T.J. Volgy. 1992. *The Forgotton Americans.* W.W. Norton and Co.,New York.

Seltzer, C. 1986. The Future of Coal and Coal Field Development: Observations and Prescriptions. In *Proceedings from the 1986 Conference on Appalachia.* Appalachian Center for the Land and Economy of Appalachia. University of Kentucky, Lexington. Oct. 30th-31st: 46-50.

Shaw, E.B. 1963. *Anglo-America: A regional Geography.* John Wiley inc., New York.

Shaw, W. 1996. *Appalacian Kentucky: Some Comments Concerning Poverty and the Nature of the Banking Network in the Region.* Paper Presented at the 92nd annual meeting of the Association of American Geographers, 9th-13th April 1996, Charlotte, N.C.

Shaw, W. 1992. "Variation in Poverty Levels within the Lower Mississippi Delta Development Region." *The Geographical Bulletin* 34 (2): 68-81.

Shaw, W. 1990a. "Post WWII Changes in Arkansas Migration." *The Mid-south Geographer,* 6: 16.

Shaw, W. 1990b. *Poverty in Arkansas Since World War II.* University of Arkansas, Fayetteville, M.A. Thesis.

Sheppard, E. 1990. "Ecological Analysis of the 'Urban Underclass': Commentary Hughes, Kasarda, O'Reagan and Wiseman." *Urban Geography* 11 (3): 285-297.

Singh, V.P. 1991. "The Underclass in the United States: Some Correlates of Economic Change." *Sociological Inquiry* 61 (4): 505-521.

Slesnick, D.T. 1993. "Gaining Ground: Poverty in the Postwar United States." *Journal of Political Economy* 101 (1): 1-38.

Smith, D.M. 1972. "Towards a Geography of Social Well-being: Inter-state Variations in the United States." *Antipode Monographs in Social Geography* 1, Geographical Perspectives on American Poverty: 5-16.

Smith, D.M. 1973. *The Geography of Social Well-being in the U.S.: An Introduction to Territorial Social Indicators*. McGraw-Hill, New York.

Smith, D.M. 1987. *Geography, Inequality and Society*. Cambridge University Press, Cambridge.

Smith, L. 1990. "The Face of Rural Poverty." *Fortune* (Dec 31st): 100-110.

Smith, N. 1979. "Geography, Science and Post-positivist Modes of Explanation." *Progress in Human Geography* 3: 356-383.

Smith, N. 1984. *Uneven Development*. Basil Blackwell, Oxford.

Smith, N. 1987a. "Dangers of the Empirical Turn: Some Comments on the CURS Initiative." *Antipode* 19 (1): 59-68.

Smith, N. 1987b. "Rascal Concepts, Minimalizing Discourse, and the Politics of Geography." *Environment and Planning D: Society and Space* 5 (4): 377-383.

Smith, N. 1991. *Uneven Development. Nature Capital and the Production of Space*. Basil Blackwell, Oxford.

Smolensky, E. 1981-82. "Poverty in the United States: Where do we Stand." *Focus* (University of Wisconsin-Madison, Institute for Research on Poverty), 5 (Winter): 1-11.

Solow, R.M. 1990. "Poverty and Economic Growth." *Focus*, (University of Wisconsin-Madison, Institute for Research on Poverty), 12 (3) (Spring): 3-5.

Sonstelic, J.C. and P.R. Portney 1976. Property Value Maximization as a Decision Criterion for Local Government. In *Economic Issues in Metropolitan Growth*. P.R. Portney Ed., Papers Presented at a Forum Conducted by Resources for the Future, Washington D.C., May 28-29.

Spring, J. 1985. *American Education*. Longman Inc., New York.

Stanback, T.M. and T.J Noyelle 1982. *Cities in Transition*. Allanheld, Osmun and Co., Totowa, N.J.

Stokes, R.G. and A.B. Anderson. 1990. "Disarticulation and Human Welfare in Less Developed Countries." *American Sociological Review* 55: 63-74.

Storper, M. 1987. "The Post-enlightenment Challenge to Marxist Urban Studies." *Environment and Planning D: Society and Space* 5 (4): 418-426.

Szakos, J. 1986. They're Not All Sitting Back and Taking it: Fighting for Change in Eastern Kentucky. In *Proceedings from the 1986 Conference on Appalachia*. Appalachian Center for the Land and Economy of Appalachia. University of Kentucky, Lexington. Oct. 30th-31st: 91-96.

Tamblyn, L.R. 1973. *Inequality. A Portrait of Rural America*. Rural Education Association, Washington D.C.

Task Force on Economic Growth and Opportunity. 1969. *Rural Poverty and Regional Progress in an Urban Society*. Chamber of Commerce of the U.S., Task Force on Economic Growth and Opportunity, Forth Report

Tata, R.J. and Schultz, R.R. 1988. "World Variation in Human Welfare: A New Index of Development Status." *Annals of the Association of American Geographers*, 78 (4): 580-593 December.

Thomas, R.J. 1982. Citizenship and Gender in Work Organization: Some Considerations for Theories of the Labor Process. In *Marxist Inquiries: Studies of Labor, Class, and States*. M. Burawoy and T. Skocpol Editors, University of Chicago Press, Chicago: 86-112.

Thompson, G.L. 1972. "The Spatial Convergence of Environmental and Demographic Variables in Poverty Landscapes." *The Southeastern Geographer*, 12: 14-22.

Thrift, N. 1986. Little Games and Big Stories: Accounting for the Practice of Personality and Politics in the 1945 General Elections. In *Politics, Geography and Social Stratification*, K. Hoggart and E. Kofman Editors, Croom Helm, London.

Thrift, N. 1987. "No Perfect Symmetry." *Environment and Planning D: Society and Space* 5 (4): 400-407.

Thurow, L.C. 1982. Equity, Efficiency, Social Justice and
   Redistribution. In *The Political Economy of the United States*.
   Ed C. Stotfaes. North-Holland Pub. Co., New York.

Tienda, M. 1989. "Puerto Ricans and the Underclass Debate." *The
   Annals of the American Academy of Political and Social
   Science* 501: 105-119.

Tienda, M., S.A. Smith, and V. Ortiz. 1987. "Industrial Restructuring,
   Gender Segregation, and Sex Differences in Earnings."
   American Sociological Review 52: 195-210.

Tigges, L 1992 *Structured Inequality*. Class Notes, unpublished

Tobin, J. 1990. "The Poverty Problem: 1964-1989." *Focus* (University
   of Wisconsin-Madison, Institute for Research on Poverty), 12
   (3) (Spring): 6-7.

Tootle, D.M. 1989. *Public Assistance and Economic Well-being: A
   Study of Spatial Variation Across Local Labor Markets in the
   United States*. Ph.D. Dissertation, University of Georgia,
   Athens.

Turner, J.H. and C.E. Starnes. 1976. *Inequality: Privilege and Poverty
   in America*. Goodyear Pub. Co. Inc., California.

U.S. Bureau of the Census. 1982. *Alternative Methods for Valuing
   Selected In-kind Transfer Benefits and Measuring Their Effects
   on Poverty*. Technical Paper.

U.S. Department of Commerce. 1952. *County and City Data Book*.
   U.S. Bureau of the Census.

U.S. Department of Commerce. 1962. *County and City Data Book*.
   U.S. Bureau of the Census.

U.S. Department of Commerce. 1972. *County and City Data Book*.
   U.S. Bureau of the Census.

U.S. Department of Commerce. 1980. *Census of the Population*. U.S.
   Bureau of the Census.

U.S. Department of Commerce. 1983. *County and City Data Book*.
   U.S. Bureau of the Census.

U.S. Department of Commerce. 1988. *Poverty in the United States:
   1987*. Bureau of the Census, Current Population Reports
   Series P-60, No.163, June.

U.S. Department of Commerce. 1991. *Poverty in the United States:
   1988 and 1989*. Bureau of the Census, Current Population
   Reports Series P-60, No.171, June.

U.S. Department of Commerce. 1992a. *Poverty in the United States: 1991*. Bureau of the Census, Current Population Reports Series P-60, No.181, June.

U.S. Department of Commerce. 1992b. *Statistical Abstract of the United States 1992*. U.S. Department of Commerce, Bureau of the Census, Washington D.C.

Valentine, C.A. 1968. *Culture and Poverty. Critique and Counter-proposals*. University of Chicago Press, Chicago.

Vance, R.B. 1972. How Much Better Will the Better World Be. In *Appalachia in the Sixties. Decade of Reawakening*. Eds. D. Walls and J.B. Stephenson. The University of Kentucky Press, Lexington, KY: 38-44.

Velasquez,M. 1987. Poverty in America. In *Poverty and Social Justice: Critical Perspectives* . Ed F. Jimenez. Bilingual Press, Tempe, Arizona: 13-26.

Vitelli, V.A. 1968. *Poverty and Related Factors in the Ozark Region*. University of Arkansas, M.S. Thesis.

Wachtel, H.M. 1974. Looking at Poverty from Radical, Conservative, and Liberal Perspectives. In *The Poverty Establishment*. Ed. P. Roby. Prentice-Hall, Inc., Englewood Cliffs, New Jersey: 180-190.

Wacquant, L.J.D. and W.J. Wilson. 1989. "The Cost of Racial and Class Exclusion in the Inner City." *Annals of the Academy of Political and Social Science* 501: 8-25.

Wager, P.W. Local Government. In *The Southern Appalachian Region. A Survey*. Ed. T.R. Ford. University of Kentucky Press, Lexington: 151-168.

Wallerstein, I. 1974. *The Modern World System*. Academic Press, New York.

Wallerstein, I. 1979. *The Capitalist World Economy*. Cambridge University Press, Cambridge.

Warf. B.L. 1985. *Regional Transformation and Everyday Life: Social Theory and Washington Lumber Production*. Ph.D. Dissertation, University of Washington, Seattle.

Wells, J.C. 1977. *Poverty Amidst Riches: Why People are Poor In Appalachia*. Ph.D. Thesis, Rutgers University, New Brunswick, New Jersey.

Wells, J.C. 1986. Organized Labor in Central Appalachia. In *Proceedings from the 1986 Conference on Appalachia*. Appalachian Center for the Land and Economy of Appalachia. University of Kentucky, Lexington. Oct. 30th-31st: 123-129.

Williams, B. 1992. "Poverty Among African Americans in the Urban United States." *Human Organization* 51 (2): 164-174.

Williams, D.R. 1991. "Structural Change and the Aggregate Poverty Rate." *Demography*, 28 (2): 323-332 May.

Wilson, W.J. 1980. *The Declining Significance of Race*. University of Chicago Press, Chicago.

Wilson, W.J. 1987. *The Truly Disadvantaged: The Inner City, the Underclass, and Public Policy*. University of Chicago Press, Chicago.

Wilson, W.J. 1989. "The Underclass: Issues, Perspectives, and Public Policy." *The Annals of the American Academy of Political and Social Science* 501: 182-192.

Wohlenberg, E.H. 1970. *The Geography of Poverty in the U.S.; A Spatial Study of the Nation's Poor*. University of Washington, Ph.D. Thesis.

Wolch, J.R. 1989. The Shadow State: Transformations in the Voluntary Sector. In *The Power of Geography. How Territory Shapes Social Life*. Eds. J. Wolch and M. Dear. Unwin Hyman, Boston, MA: 197-221.

Wolfe, A. 1981. *America's Impasse. The Rise and Fall of the Politics of Growth*. Pantheon Books, New York

Wolff, R. and S. Resnick. 1987. *Economics: Marxian versus Neoclassical*. Johns Hopkins University Press.

Yapa, L.S. and Zelinsky, W. 1989. "How Not to Study the Geography of Human Welfare." *Annals of the Association of American Geographers*, 79 (4): 609-611 December.

Zhou, B. and W. Shaw. 1993. *Spatial Variation in Cost Efficiency, Optimal Size, and Output Quality in School Districts: Evidence from Georgia*. Unpublished paper.

Zimbalist, S. 1964. "Drawing the Poverty Line." *Social Work* 9: 19-26.

Index

Absentee Landownership,
    155-156, 199
Absolute Poverty, 31
Adjusted Per Capita Income
    See INDEX
AFDC, 33-34, 152
Affluence, 3, 31, 187
Affluent Counties, 174-176
African American
    See also Black-white
    Duality; Mississippi
    Delta Poverty Core
    children, 57
    discrimination, 15, 146,
    149, 151-152
    education, 55, 59,
    151-152
    and illegitimacy, 58
    infant mortality, 54, 59
    occupations, 15
    poverty levels, 58-59,
    100-101, 148
    rural poverty, 142, 145,
    146-147, 149, 151-152
    segregation, 15
    southern, 64, 149,
    151-152
    and unions, 15
    urban poverty, 145,
    140-141
Age Structure, 85, 98, 100, 111,
    129, 139, 142, 145,
    153-154, 181, 187-188

See also Family
    Structure
Alaska, 8
Alcoholism, 63, 163, 193
    See also Health Care
Apache County, AZ, 74
Appalachian Poverty Core, 5, 65,
    75, 79-80, 148, 155-156,
    159-160, 163, 164, 174,
    185, 190, 197-199
    See also European-
    american
Appalachian Regional
    Commission, 198
Asian Americans, 37
Assets, 45, 47
Automation, 19, 156

Banking, 198
Black
    See African American
Black-white Duality, 94, 100,
    109, 127, 139, 142-143,
    148, 181, 185, 190
    See also African
    American
Blue-collar Economy
    See Manufacturing
    Economy
Bronx County, NY, 72
Business Cycle, 3, 10

Capital Intensive Industry, 159

See also Manufacturing

Capitalism, 14-15, 22, 155-156, 163, 190, 193, 198-199

Catron County, NM, 72

Central City
    See Inner City

Characteristics of the Poor, 56-60
    See also Poor

Childcare, 191-192

Children, 4, 56-58, 101-102, 153, 191, 196
    See also Family
    Structure

Class Struggle, 22
    See also Capitalism

Clarke County, GA, 152

Clay County, WV, 72

Coal Industry, 155-156, 163, 199
    See also Appalachian
    Poverty Core

College Education, 153

Communality Estimates, 94, 96, 179
    See also Principal
    Components Analysis

Commuting, 83-84, 99-100, 102, 120, 138, 141-142, 164-165, 182

Comparative Advantage, 11

Conservative View, 17, 23-26
    See also Poverty Theory

Consumer Price Index, 41, 43

Consumption, 47

Core Sector, 20-21

Correlation
    analysis, 93
    between variables, 87-91

Corruption, 198-199

Corson County, SD, 74

Costilla County, CO, 72

Cost of Living, 44

Cumberland County, KY, 72

Dakotan Poverty Core, 74, 79-80, 153-155, 172, 185, 190, 195-196
    See also Native American

Data, 8, 67-69, 81-87, 88, 168-169, 171
    See also Variables

Decision Making, 12

Denver, CO, 153

Deserving Poor, 16, 26-28

Development, 10-11

Dimensions of Poverty, 53-56
    See also Poverty

Discrimination
    See also Racism
    age, 13
    ethnicity, 146-147, 162
    gender, 13-15
    institutionalized, 3, 16, 190, 196-197
    race, 13, 15, 146-147, 190, 196-197

Drug Abuse, 63
    See also Alcoholism;
    Health Care

Dual Economy, 20-21

Dummy Variables, 125
    See also Regression
    Modeling

East Coast, 79

Eastern Kentucky
    See Appalachian Poverty
    Core

East North Central Region, 148, 150, 153, 160, 164, 167

East South Central Region,

148-149, 157, 160, 161,
164, 167
Economic Opportunity Act, 41
Economic Structure
    and African Americans,
    27-28
    capitalism,4, 14-15, 22
    flexible specialization, 4,
    19
    global, 4, 21, 26
    manufacturing, 4, 18
    mass production, 4, 19
    occupations, 4
    restructuring, 26-28
    sectors, 4, 20-21
    services, 4, 18-19
Economic System, 4, 14-16, 21,
22
Economic System Theories, 18-
22
Economy Food Plan, 41-42
Education
    See also Schools
    African Americans, 25,
    55, 151-152, 197
    funding, 54, 151-152
    insufficient, 16-17,
    54-55, 183, 187
    Native Americans,
    193-194, 196
    quality, 54, 145-146,
    187, 199
    reform, 197
    rural, 55
    post high school, 55,
    153
    variables, 81
Educational Level, 181, 182-183,
190
Eigenvalue, 87, 93, 95, 179

    See also Principal
    Components Analysis
Elderly, 15-16, 56-57, 101-102,
    187-188
Engel's Coefficient, 41, 44
Employment
    See Unemployment;
    Work
Employment Status, 98, 100,
    113, 131, 140, 142-143,
    155-157, 182
    See also Unemployment;
    Work
Environment, 11, 199
Environmental Determinism, 11
European American
    See also Appalachian
    Poverty Core
    and illegitimacy
    poverty levels, 37, 58-59
    work, 19-20
Expansion Method, 108, 123-124,
    169, 171
    See also Regression
    Modeling
Exploitation, 22, 155-156, 161,
    195, 199
    See also Capitalism
External Labor Market, 20-21
Extreme Age Population
    See Age Structure

Factor Loadings, 92-99
    See also Principal
    Components Analysis
Factors, 87, 93-99, 17 -182
    See also Principal
    Components Analysis
Family Structure
    See also Children;

Elderly; Female Headed
Households; Illegitimate
Births
children, 153-154, 164,
196
and poverty line, 42
variables, 85
Faulk County, SD, 170
Female Headed Households, 94,
101
See also Children;
Elderly; Illegitimate
Births
Females
See Women
Fetal Alcohol Syndrome, 163
See also Alcoholism
Flawed Character, 12
See also Poverty Theory
Food Costs, 63
Food Plans, 41
Food Stamps, 45, 64
Free Market, 11, 20

Gaming, 185
See also Native
American
Gender, 57-58, 85
See also Women
Genetic Deficiency, 12
Geographic Divisions, 108, 122,
125
Georgia, 151-152
Ghettoization, 147
Global Economic System, 21
Government Policy, 3, 16-18, 40
Government Responsibility
Theories, 18
Great Depression, 40
Group Responsibility Theory,

12-13
Growth
See Population Change
Growth Centers, 198

Harrington, Michael, 27-28
Hawaii, 8
Health, 53-54, 63
See also Health Care
Health Care
availability, 99-100,
103-104, 118, 136, 141,
143, 146, 162-163, 191
insufficient, 16, 53-54,
63
reform, 104, 191, 193,
196
rural, 146
variables, 83
Health Service Economy, 98,
100, 102-103, 115, 133,
140, 142-143, 145,
159-160, 195
High-tech Industry, 159
See also Manufacturing
Hispanic Americans, 37, 57, 74,
100-101
See also Spanish Culture
and Ethnicity; Texas
Border Poverty Core;
Hoover, Herbert, 31
Hospitals, 54, 146, 163
See also Health Care
Housing
costs, 47, 56, 68-69
quality, 55-56, 63,
181-182
overcrowding, 56,
181-182
variables, 81

Hudspeth County, TX, 72
Human Capital
    See Education

Illegal Immigrants, 157, 194-195
Illegitimate Births, 58
    See also Children;
    Family Structure;
Female Headed Households
Incentives, 158
Income Distribution, 45
    See also Inequality;
    Poverty
Income Index
    See INDEX
Independent Variables, 92
    See also Regression
    Modeling
Indian
    See Native American
Individual Responsibility Theory,
    12-13
INDEX, 69-72, 92, 172-173
Inequality, 35, 45, 178, 197
    See also Income
    Distribution; Poverty
Infant Mortality, 54, 59, 63, 81,
    83, 163
    See also Health Care
Informal Economic Activity, 146
Infrastructure, 147
    See also Commuting;
    Transportation
Inner City, 61-62
    See also Urban Poverty
Institutional Responsibility
    Theories, 16-18
Institutions, 16-18, 151, 193
Internal Labor Market, 20-21
International Division of Labor,

21
    See also Manufacturing

Jobs
    See Unemployment;
    Work
Johnson, Lyndon B., 31-33, 41
    See also War on Poverty

Labor Cost, 145, 155, 158,
    194-195
    See also Manufacturing
Labor Force, 86
    See also Unemployment;
    Work
Labor Intensive Industry, 158,
    192, 198
    See also Manufacturing
Labor Mobility
    See Commuting
Land Ownership, 197
    See also Absentee
    Landownership
Laziness, 12, 17, 23-26
Lewis, Oscar, 13
Liberal Food Plan, 41
Liberal View, 17, 26-28
    See also Poverty Theory
Local Development Districts, 198
Low Cost Food Plan, 41
Lumber Industry
    See Timber Industry

Manufacturing
    decline, 18-19, 21, 60,
    158
    flexible specialization,
    4,19
    mass production, 4, 19
    nature of, 139, 158-159,

197
Manufacturing Economy, 98,
      100, 102-103, 114, 132,
      140, 142-143, 145,
      157-160, 192-193, 195
Marital Status, 57-58
      See also Family
      Structure; Female Headed
      Households
Market Baskets, 40-45
Maverick County, TX, 74
Median Income, 44-46
Medicaid, 35
Medicare, 16
Men
      dominance by, 14, 19-20
Methodology, 70, 87, 92-93,
      168-169
Mexico, 157, 194-195
Middle Atlantic Region, 150,
      160, 167-168
Migration, 85-86, 157, 164
      See also Illegal
      Immigrants
Miner County, SD, 172
Mining
      See Coal Industry
Mississippi Delta Poverty Core,
      5, 62-63, 65, 74-75,
      79-80, 148, 172, 174,
      187, 190, 196-197
      See also African-
      american
Moderate Cost Food Plan, 41
Mormons, 154
Mountain Region, 149, 152-153,
      155-156, 157, 161, 163,
      164, 166
Multinational Corporations, 21
Murray, Charles, 24-26

NAFTA, 195
Native American
      culture and ethnicity,
      99-100, 117, 135, 141,
      142-143, 154, 161-163,
      181, 182, 192-194
      poverty, 65, 75,
      100-101, 142-143,
      146-147, 153-154, 185,
      192-194
      reservations, 153, 159,
      161-162, 192-194
      See also Dakotan
      Poverty Core;
      Southwestern Poverty
      Core
Navajo, 162-163, 193
New Deal, 40
New England Region, 150, 168
New International Division of
      Labor, 21
No Fault Theory, 10-11
Non-income Poverty, 46-49
Nonmetropolitan
      See Rural
Non-yuppieness. 94, 98-99, 110,
      128, 139, 142, 145,
      152-153, 192-193
Northeast, 158
Nutritional Needs, 41-42

Occupational Segregation, 4,
      14-15, 19-20
Oklahoma, 75
Organized Labor, 15, 19, 22, 158,
      163
Orshansky, Mollie, 42-45
Overcrowding, 56
Ozarks, 65, 75

Pacific Region, 149, 157, 166
Parameter Estimates, 99-100,
    144, 147-150, 182-184,
    243-250
    See also Regression
    Modeling
Patriarchal System, 14
    See also Men
'Peach' Training Program
    See Training Programs
Per Capita Income
    1980s, 70, 72-74
    1990s, 36, 172, 174,
    169
    as a poverty measure, 49,
    68
Periphery Sector, 20-21
Persecution, 161
Policy, 191-199
Political Climate, 5-6
Politics, 198-199
    See also Power
Poor
    African American, 58-59
    behavior, 12-13, 17,
    23-28
    characteristics, 4, 23-28,
    56-60
    children, 4, 23-26, 57-58
    decision making, 12,
    23-26
    education, 16, 23-28,
    54-55
    elderly, 15-16, 56-57
    European American,
    58-59
    human capital, 16, 23-28
    location, 4, 67-81,
    172-175, 178
    stereotypes, 13, 17,

    23-26
    women, 4, 13-15, 57-58
    working, 4, 60, 102
Poorest Counties
    1980, 71, 73, 77, 172,
    175-176, 178
    1990, 170, 172, 173,
    174, 175-176, 178,
    237-241
Population Change, 11, 99-100,
    104, 119, 137, 141, 143,
    145, 164
    See also Migration
Population Growth
    See Population Change
Possibilism, 11
Poverty
    African Americans and,
    13, 23-28, 37, 58-59
    causes of, 3, 9-28
    change over time, 5-6
    and children, 4, 23-26,
    57, 101-102
    definition of, 38-51
    density, 70
    dimensions of, 53-56
    distribution, 67-81
    European Americans and,
    58-59
    gender, 57-58
    See also women
    geography, 4, 6, 64-65
    and education, 16, 54-55
    and the elderly, 15-16,
    56-57, 101-102
    and family, 12
    and government, 3, 10
    and health care, 16,
    53-54
    and housing, 55-56

and individual choices, 3, 6, 10, 23-26
and institutions, 3, 10, 16-18
inter-generational, 13
lines, 39-49
measurement of, 38-51
national levels, 3
national model, 92, 99-104
non-income, 46-49
persistence of, 3
and power, 10, 17
and race
See Race
regional variation in, 125-168, 189-190
regions, 4, 64-65
rural, 4-6, 37, 59, 61-64, 123-125, 142-143, 155
and society, 3, 26-28
spatially varying model, 107-168
temporally varying model, 168-188
and unemployment, 59-60
urban, 4-5, 23-28, 37, 61-62, 123-125, 142, 143-147
urban-rural model, 123-125, 142-147
and women, 13-15, 36-37, 57-58
Poverty Cores, 67-81, 172-175, 178, 189
Poverty Definition, 38-51
Poverty Density, 70, 78-79
Poverty Levels
1960s, 33
1970s, 33
1980s, 33-34, 168-188
1990s, 168-188
contemporary, 35-38
historical, 31-35
post-industrial, 32
Poverty Lines
Great Britain, 40
historical, 39-40
official, 3, 32, 41-45
Orshansky system, 42-45
rural, 42-43
Poverty Measurement, 38-51, 67-69
Poverty Theory
blaming the victim, 9, 12-13, 17, 23-26
blaming the system, 9, 13-16, 26-28
control and power, 10, 17
conservative, 10, 17, 23-26
culture of poverty, 10, 12-13
government, 10, 12-13
institutions, 10, 16-18
liberal, 10, 17, 26-28
no fault theory, 10-11
radical, 10, 22
and resources, 10-11
shortcomings of 28-29
structural, 10, 18-22, 26-28, 145
Power
and poverty, 10, 17, 22, 194, 198-199
Presido County, TX, 74
Prices, 63
Primary Labor Market, 20-21

Principal Components Analysis
    87, 92-99, 169, 178-182

Race, 58-59, 84-85, 100-101,
    146-147, 182-183
    See also African-
    american; Asian-
    american; European-
    american; Hispanic-
    american
Racism, 13, 15, 19-20, 149,
    151-152, 160, 196-197
    See also African-
    american
Reagan, Ronald, 35
Recommended Dietary
    Allowances, 41
Reconstruction Era, 151
Regions, 64-65
Regression Modeling, 92-93,
    99-104, 142, 147-148,
    171, 172-183
Relative Poverty, 31, 40, 46-47
Research Questions, 7
Reservations
    See Native American
Residential Status, 84
Resource Endowments, 3, 10-11
Rich County, UT, 72
Robotization, 19
Rural Counties, 121, 147, 162
Rural Poverty, 4-6, 37, 59,
    61-64, 108, 123-125,
    142-143, 149, 151-152,
    155, 189-190, 198

Sanborn County, SD, 172
Schools, 63, 151-152
    See also Education
Seasonal Employment, 146

Secondary Labor Market, 20-21
Segregation
    occupational, 4, 14-15,
    19-20. 152
    racial, 15, 148, 151
Service Sector, 18-19, 60, 145
Sexism, 13-15, 19-20
Shannon County, SD, 74, 153,
    187, 196
Silicon Valley, 159
Social Deviance, 12-13, 23-26
Social Networks, 146, 151
Social Policy
    See Policy
Social Security, 16, 57, 187, 191
Social Spending, 16-17, 57
Societal Responsibility Theory,
    13-16, 26-28
Society
    attitudes, 13, 151
    culture, 13, 151
    economic action, 13
    values, 151
South, 37, 55, 64-65, 148, 149,
    158-159, 160, 190
South Atlantic Region, 149, 157,
    165-166
Southern Coastal Plain, 5, 65
    See also Mississippi
    Delta Poverty Core
Southwestern Poverty Core, 74,
    79-80, 153-154,
    161-162, 164, 172, 185,
    190, 192-194
    See also Native American
Spanish Culture and Ethnicity,
    99-100, 116, 134,
    140-142, 146-147, 160,
    181, 182, 194-195
    See also Texas Border

Poverty Core
Spatially Varying Model, 107-168
Starr County, TX, 72, 74, 172
State, The, 16, 21
Stepwise Regression, 124-126,
        142, 147, 171
Structural View, 26-28
        See also Poverty Theory
Subsistence Level, 39-40
Suburbs, 61

Taxes Paid, 45, 56, 158, 199
Tax System, 18, 35, 56, 158, 199
Teacher Attitudes, 152
Texas, 65, 74
        See also Texas Border
Poverty Core
Texas Border Poverty Core, 74,
        79-80, 156, 160, 172,
        185, 190, 194-195
        See also Hispanic-
        americans
Theory, 3, 9-29, 190-191,
        191-199, 200
        See also Poverty Theory
Thrifty Food Plan, 42
Timber Industry, 155-156, 163,
        199
Todd County, SD, 74
Tourism, 196
Training Programs, 152
Transfer Payments, 16, 31, 57,
        63-64, 85, 187, 191
Transportation, 165, 196
        See also Commuting
Trickle Down, 11
Tunica County, MS, 62-63, 172

Undeserving Poor, 16, 23-26
Unemployment, 10, 21-28,

        140-141, 145-146,
        155-157, 183, 185-187,
        190, 194-196, 197
Unions
        See Organized labor
Urban Counties, 121, 147
Urban Poverty, 61-62, 108,
        123-125, 189-190
Urban Residential Status,
        182-183
Urban Retail Economy, 98, 100,
        103, 112, 130, 139-140,
        142-143, 145, 154-155,
        195
Urban Underclass, 4, 23-28, 61-62
USDA Food Plans, 41-45
Utah, 154

Values, 24-26, 31
Van Buren County, TN, 72
Variables, 81-87, 169, 179
        See also Data
        relationship with
        INDEX, 93
Violence, 163

Wages, 10, 18-19, 60, 63, 85, 98,
        194-195, 198
        See also Labor Costs
War on Poverty, 33, 38, 41
War on Welfare, 33
Welfare Programs, 4-6, 16-17,
        25-26, 145, 152
        See also Transfer
        Payments
West, 159
West Coast, 79
        See also Pacific Region;
        West
West North Central Region, 149,

161, 167

West South Central Region, 149,
  155-156, 166

West Virginia, 174
  See also Appalachian
  Poverty Core

White
  See European American

Wilson, William Julius, 27-28,
  58, 185
  See also Urban
  Underclass

Women, 4, 13-15
  See also Family
  Structure; Female Headed
  Households; Gender

Work
  See also Unemployment
  and African Americans,
  15, 19-20, 84-85, 152
  in agriculture, 146, 156
  and European Americans,
  19-20
  and Hispanic Americans,
  194-195
  location of, 18, 147
  in mining, 156
  opportunities, 13, 15,
  19-20
  searching for, 13, 19-20
  seasonal, 146
  security, 19, 21-22
  in tourist industry, 146
  and women, 14, 19-20

Work Ethic, 12, 17, 191-192

Working Poor, 187, 191-192

Yuppies
  See Non-yuppieness

Zavala County, TX, 74

Ziebach County, SD, 153